Free Space!

Real Alternatives for Reaching Outer Space

by B. Alexander Howerton

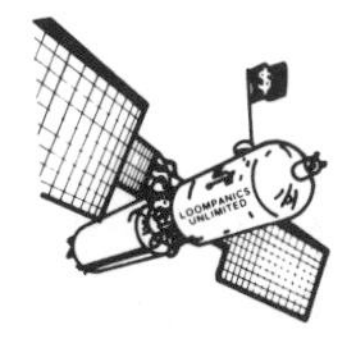

Loompanics Unlimited
Port Townsend, Washington

This book is sold for information purposes only. Neither the author nor the publisher will be held accountable for the use or misuse of the information contained in this book.

FREE SPACE! Real Alternatives for Reaching Outer Space

Cover design by Dan Wend/MediaGRAPHICS
Cover Photo: The proposed base camp for the Mars Direct project could be operational as early as 2003. Photo courtesy of Robert Murray, Martin Marietta Astronautics.

Published by:
Loompanics Unlimited
P.O. Box 1197
Port Townsend, WA 98368

Loompanics Unlimited is a division of Loompanics Enterprises, Inc.

ISBN 1-55950-120-0
Library of Congress Catalog Card Number 94-79612

Contents

Part I
The Problem

Part II
The Solutions

To my father, Beverly R. Howerton,
who has endless patience and faith in me.

Soar

Part I
The Problem

One
NASA:
"The Eagle Has Landed..."
And Has Remained Stuck On The Ground

The course of humanity was forever changed on July 20, 1969. Many observers have likened the first Moon landing to the day when Columbus first set foot in the "New World," but the significance runs much deeper. That first tentative yet indelible footfall on another planet corresponds to that undateable day when the first fish finally succeeded in breathing outside the protective envelope of the ocean. We are attempting something no less challenging, for we must take our protective envelope with us into space, or we perish. When we imagine how long it must have taken, since the dawn of life on Earth, for that fateful migration from the ocean to come to pass, how many individual fish died in the process, how many dead-ends and how much backsliding there must have been, it is truly remarkable that we have done so much in space in so short a time, with so few accidents and casualties.

Free Space!

4

Let us give the Apollo program its proper respect. The Moon landing was, very simply, the most amazing thing humans have ever accomplished. It was an incredible technical feat. In eight short years, from President Kennedy's challenge, to Apollo 11, we progressed from not even knowing if humans could survive for long periods outside the atmosphere, to landing a man on the Moon and returning him safely to Earth. It was a stupendous cultural achievement. Neil Armstrong was not alone when he stepped onto the Moon; thousands of people spent many years to build the awesome machines that took him and his crewmates there and back, and we all went with them through the marvels of television. It was an amazing political achievement. The Soviet Union and the United States decided to engage in a marathon to the Moon, in which the best contender would win, rather than indulging in the boxing match of Mutually Assured Destruction, in which the victor stands over his vanquished opponent, triumphant, yet as battered and bloody as his foe. The Apollo program was nothing less than the best humanity *could* achieve.

Yet in the decades since that glorious achievement we have retreated from the Moon, to the point that we can no longer return. A joke current among space enthusiasts goes, "If we can send a man to the Moon, why can't we send a man to the Moon?" In these politically correct times, the joke also has a variant: "If we can send a man to the Moon, why can't we send a woman?" Jokes often indicate an incisive underlying perception of reality, and these quips are no exception. We have retreated from the glory of Apollo because the program was founded on the shifting sands of political fortune. Moreover, the benefits of Apollo were never shared with the greater populace, which ultimately supported the program with tax dollars and enthusiasm. The exploits of a few fly-boys with "the right stuff" can only entertain for so long. These days our remote control fingers are much faster than the best reflexes of an Apollo astronaut. The average person realized she or he was never going to be invited along for the ride, and our attention turned elsewhere.

The seeds of the decline of NASA were sown during its inception. The National Aeronautics and Space Act of 1958, which created NASA, was a direct reaction to the Soviet launching of Sputnik I. The advance into space was a military race from day one. Opinions expressed by men in power during the Kennedy administration reveal the extent of the fear which goaded them. Donald G. Brennan, an arms control advisor to President Kennedy, wrote:

> It is worth stressing at the outset that the competition in space technology generally, and in its military applications specifically, is one of the aspects of the cold war and cannot be divorced from that setting. As such, it requires an understanding both of the political setting imposed by the cold war and of the framework of military strategy within which the cold war in its military aspects is being carried on — and may, unless checked, be carried on in space.
>
> Although Soviet officials have not said much explicitly about their present and projected military capabilities in space, they possess a demonstrated prowess in basic space technology which in the context of cold war rivalries carries with it an implied danger of military applications. This implied threat casts a shadow on the international political scene — as, of course, does our own military power and progress in space — especially since, with the present relative military standoff, space offers some chance of achieving a decisive military advantage over one's

> opponent. Insofar as either side convinces itself of the possibility of scoring a strategic gain by extending military competition into space, it will be under pressure to do so.[1]

A more hotheaded approach was advocated by William V. Shannon, a columnist at the *New York Post*:

> Fate has made us the trustee of mankind's freedom. We are engaged in a desperate, hand-to-hand struggle with the Communists who have already captured control of the great peoples of Russia and China and who want to conquer and enslave all the rest of us. Anything that nourishes the strength of the Soviet ruling clique and enhances its prestige is a misfortune for the cause of freedom....
>
> The news of the Soviet space triumph is therefore as tragic an event as if during World War II Hitler's scientists had gone ahead of us in the race to develop radar or the atomic bomb. Science, if divorced from moral purpose, only makes tyranny more efficient. Winston Churchill declared on the day France fell in 1940:
>
> "If we can stand up to [Hitler], all Europe may be free and the life of the world may move forward into broad, sunlit uplands. But if we fail, then the whole world, including the United States, including all that we have known and cared for, will sink into the abyss of a new dark age made more sinister, and perhaps more protracted, by the lights of perverted science."
>
> We need only substitute ourselves for the British and Khrushchev for Hitler in these sentences to find

> that Churchill has described our present situation exactly.[2]

It thus became our "moral purpose" to beat the Soviets into space. All subsequent space activities occurred within this paradigm. In fact, it became such an overwhelming goal that once we had done it, we threw away our space capability. The last three Apollo missions were canceled by President Johnson, and President Nixon took so long in approving a start for the Space Shuttle that its first flight occurred in 1981, two years after Skylab, its original destination, had fallen back to Earth. Yet by then space was so inextricably linked with NASA in the minds of the American people, that they had faith that NASA would one day open up the heavens for all of us. G. Harry Stine's 1979 books *The Third Industrial Revolution* and *The Space Enterprise* are brimming with the hope and the promise the Space Shuttle held for commercial space. And when President Reagan announced the commencement of Space Station Freedom in 1984, everyone was convinced that routine travel to space was just around the corner.

Unfortunately, a fateful day in January of 1986 dashed those hopes to the ground. The Space Shuttle Challenger exploded, and with it our dreams of routine space travel also blew up. NASA decided that the Space Shuttle was still an experimental vehicle, and only astronauts would ever ride on it. The Space Shuttle returned to service in late 1988, but is still as feisty and problematic as a thoroughbred, and the plans for the Space Station keep slipping, and slipping, and slipping...

In a very real sense, the Challenger incident closed the chapter on space exploration that the first Moon landing opened. During that period, the American people thought that the Space Program was proceeding nicely, and that the government was the proper organization in which to invest our hopes of moving off this world. NASA, that glorious monument to technical excellence that brought us the Moon landings, could do no wrong, and would take us all to the stars. Challenger changed all that. Challenger was not only indicative of the hubris that had set in at NASA, which would have manifested itself eventually in some sort of disaster; it was also the beginning of the long, slow slide from which NASA has never recovered.

We all wish that the Challenger disaster had never occurred. Yet, given the circumstances and the cultural climate, it had the grim inevitability of a Greek tragedy. We were acting out the parts we had written for ourselves in the early 1960s, with the unstoppable power of the self-fulfilling prophecies that guided Oedipus to his doom. We had declared the advance into space to be a military race, then pursued our course with martial vigor. Our astronauts were considered soldiers of a sort, even once we began admitting civilians into the corps. Although mission success is critical in military operations and in activities modeled on them, ultimately the soldiers are as expendable as the materiel in the thrust for greater glory. That is why the safety requirements were relaxed that January morning. That is why we have dead heroes to mourn.

And mourn we did, not only for the lost lives, but also for the lost glory, for the end of innocence, for the death of the promise of Apollo. NASA got the Shuttle flying again in under three years, but ever since, the esteem and prestige of NASA have been steadily eroding. Moreover, since space is inextricably linked with NASA in the American mind, space in general is suffering a loss of enthusiasm and morale.

NASA is currently stuck in a pit that it dug for itself. One disaster after another has pocked NASA's face. Of course, there have also been successes, but those are of little consequence to a public that feels betrayed. NASA cannot seem to do anything right anymore. The Hubble Space Telescope was launched with a flawed mirror. The Galileo spacecraft was unable to unfold its high-gain antenna. A fuel hose on Mars Observer exploded, and the craft was lost forever. An experimental tether failed to unreel properly. The Shuttle keeps getting grounded. And the plans for the space station keep slipping, and slipping, and slipping...

This is the inevitable result of a monopoly, which is exactly what NASA is. The agency which once facilitated our trip to the Moon is now impeding our progress into the heavens. All space plans bottleneck through NASA, which has limited funds. Therefore unconventional and non-status-quo ideas get rejected as unworkable or too risky. Yet NASA's stay-the-course mentality is not getting anyone closer to going to space permanently.

The monolithic structure of NASA is yet another manifestation of the cold war paradigm. To defeat our enemy, we had to become our enemy. To tear down the monolithic system of the Soviet Union, we had to build our own, bigger monolith. We succeeded in our goal, but our own Beast is now turning on us, like Frankenstein's monster. It is the ultimate irony.

NASA is exhibiting all the symptoms of a monopoly. An October 29th, 1992 General Accounting Office (GAO) report found sloppy accounting practices throughout NASA headquarters and its four biggest field centers. According to *Aviation Week and Space Technology*, "The report said NASA's failure to enforce spending limits and keep track of property has been a problem 'for many years.' It said the GAO identified the agency's contract administration as 'high risk' in January, 1990, because it had failed to correct weaknesses in management and in-house expertise had eroded."[3] Another GAO report found that NASA programs ended up costing an average of 77% over their original estimates, due to redesigns, technical problems, budget constraints, faulty cost estimating, Shuttle delays, and inflation.[4] The Mars Observer review team, investigating the craft's loss, found "'a number of spacecraft design flaws and poor operating procedures' and criticized program management by NASA's Jet Propulsion Laboratory (JPL) in Pasadena, Calif."[5] *Space News* reported in April 1994 that Wes Huntress, NASA's own associate administrator for space science found that:

> NASA wastes money and manpower by relying on outdated technologies and attitudes to operate its spacecraft....
>
> Huntress ... said the agencies' own bureaucracy had resisted improvements that would dramatically lower operation costs....
>
> We are two decades behind the curve in engineering the way we operate our missions....
>
> There is no reason ... why we cannot run our spacecraft from a few workstations with limited science operations and with far fewer people than we are using.... We can no longer accept the dogma that so many people, all looking over one another's shoulders, are so important for assuring mission success. There are better ways to achieve success.[6]

Bill Colvin, NASA's independent inspector general, found the same types of laxity as Huntress and the GAO. He and Frank Conahan, assistant comptroller general at the GAO, in testimony before a House of Representatives subcommittee on October 6, 1993, summarized years of investigations and dozens of reports in their testimony. Conahan said his office has identified NASA contracts as one of 17 government areas at highest risk for waste, fraud and abuse. The investigators cited a host of problems, including:

> • Sloppy procurement practices that allow many contracts to be awarded without competition and enable companies to charge exorbitant fees, primarily because NASA does not independently assess the costs of its projects.
>
> • Inadequate oversight of contracts and practices that routinely pay out merit fees on contracts experiencing huge overruns and poor technical performance.
>
> • An accounting system so riddled with errors and inconsistencies that the inspector general was unable to audit the agency's books as required by federal law.
>
> • Failing to complete a strategic plan. The Senate Appropriations Committee is still waiting for the plan, which it requested two years ago.
>
> • Decentralized management that gives the administrator little control over the agency's research centers.
>
> The result is that NASA takes more money to accomplish less work than it should, Colvin said.[7]

The ultimate arrogance of any monopoly is to believe the law does not apply to it, or that it can get away with something, because it can be covered up, and there is no other supplier to turn to. This attitude is illustrated by the indictment of several Johnson Space Center employees and associated contractors. *Space News* reported on February 28, 1994, that "a 20-month FBI investigation of NASA and its contractors culminated last week in multiple criminal charges alleging misconduct in a range of activities managed by Johnson Space Center, including several multimillion-dollar contracts."[8]

Dan Goldin, NASA Administrator since early 1992, is aware of the monopolistic tendencies of NASA. Formerly a project manager at TRW, he knows how to run a business. He has a vision for NASA's future, and he knows he cannot accomplish it unless he reforms NASA. Goldin wants to move NASA and humanity out to the planets, first by completing the Space Station. Next would come a lunar outpost, where we would have an excellent vantage point for astronomy, especially in the search for planets around other suns. We could also catalog and begin to use lunar resources, and test complex space equipment in an appropriate environment. Then we would be off to Mars, robotically at first, followed by a crewed expedition. We would catalog Martian resources, but more importantly, we would look for signs of life. Within twenty years, Goldin wants to have mapped the resources of all the major bodies in the solar system, and even to have returned a few samples to Earth. He believes none of this will transpire without international cooperation, business-like reforms within NASA, building programs "faster, better, cheaper," and communicating the results to the people, who ultimately pay for and support the program.[9]

Goldin knows he will never achieve his vision if he fails to tame and train the monolithic Beast. He is a champion of Total Quality Management, which seeks to concentrate on providing customers with what they want and demand, by continually improving processes and procedures. To that end, he made sweeping management changes throughout NASA early in 1994.[10] He wants to involve business more, and decentralize the grip of NASA by spinning off technologies to the private sector. *Space News* reported that in a November 10, 1992, speech "Goldin also said the space business sector must move beyond its attempts to sell goods and services to the government to become a real commercial industry.... 'NASA wants to commercialize space, not privatize space for people who want to sell NASA goods and services.'"[11] However, Goldin's speech "was greeted with some anxiety from the audience, made up primarily of NASA contractors."[12] The Beast does not want to be tamed; NASA and its contractors have a very cozy relationship, and that is the way they want to keep it. Goldin knows that, however. When he started his reforms in 1992,

> Goldin said he expects a backlash from some agency employees and contractors against his current attempt to cut the growth in NASA budgets....
>
> "I ... hope the contractors allow NASA for the first time in many decades to take charge of the civil space program...."
>
> Moreover, Goldin views NASA as moribund. The agency is choking on its own bureaucracy and contractor support system and it is unable to quickly carry out its objectives....
>
> "A major portion of how we spend money goes into paper and not into hardware...."
>
> "I think collectively, we as the civil aerospace community has [sic] lost sight of the direction we ought to go in."[13]

Goldin's rallying cry for his reform is, "Faster, better, cheaper!" He and others point to the success of the Clementine 1 spacecraft as an example of how projects should be managed. Clementine 1 was designed, built, and launched in twenty-two months by a small team of fifty-five workers, and it came in under its $55 million budget. There is only one problem; it was primarily a Department of Defense project, and NASA had minimal involvement. "NASA — a partner in the mission with the Ballistic Missile Defense Organization — has been giving less than wholehearted support to the program."[14] The project's manager, Lieutenant Colonel Pedro Rustan, believes the mission should become the standard for faster, better, cheaper spacecraft. "The spacecraft has been designed, built, tested and controlled in space by a team of about 55 people. We don't need a lot of fancy scientist Ph.D.s to build a spacecraft."[15] NASA believes it should become the model, too, and has started a $108 million program to do what Rustan did with $55 million. The Beast refuses to be tamed. Rustan is somewhat disgusted. "The most important lesson [of Clementine 1], Rustan said, is that the government is better equipped than private industry to build demonstration spacecraft."[16] Rustan drew the correct conclusion, but attributed it to the wrong factors. The Department of Defense is quite a monolithic beast in its own right. The reason his team pulled off Clementine 1 is that they were focused, whereas NASA, despite Goldin's vision, is aimlessly wandering around over a range of programs it cannot quite manage. Any project with a tight focus, whether it be carried out by the Department of Defense, NASA, or private industry, will breed better success than random activity. Clementine 1 has since failed, but that does not detract from the achievements of the design team; it accomplished its primary mission flawlessly.

NASA has attempted "faster, better, cheaper" reform, but the clumsy Beast does not quite have the dexterity to manipulate such a fine concept. In April, 1994, the agency presented to potential bidders a 700-page request for proposals for a Mars Global Surveyor, loaded with specifications that would limit a project leader's ability to do the project faster, better, cheaper. One prospective project contractor said, "That generally means they will dink around and put more requirements on things, which ups the cost."[17] Another contractor said:

> When you look at the entire request for proposal, JPL's full and normal bureaucratic process appeared to be sitting there, ready to sit hard on us. There's a good chunk of money that would go to JPL for keeping a lot of folks employed in a supporting oversight role. Frankly, we were a little surprised that NASA headquarters didn't question that....
>
> In response to the criticism, William Piotrowski, acting director of NASA's solar system exploration division, said, "We felt that in the case of Mars Global Surveyor, JPL was not asking an inordinate amount of oversight. We feel comfortable with the level of oversight that was proposed."[18]

The Beast feels comfortable, while its "slaves" are reeling under the burden of its bulk.

Goldin is on a valiant hero's quest to conquer the monstrous Beast, but he may be too late. A Congressional Budget Office report dated March 23, 1994, states that "NASA's aggressive reform strategy will not save enough money to help the agency pay for its full plate of projects in the next few years.... NASA should consider canceling major programs, such as the space station and space shuttle."[19] The current budget battle in Congress over the fate of the Space Station is starting to be perceived as a referendum on the future of NASA:

> "The stage is finally set for a debate on what kind of space program we want," said John Logsdon, [director of the Space Policy Institute] at George Washington University. Adds a congressional staffer, "You are witnessing NASA realigning itself as an institution, not just in the programs...."
>
> In the long run, NASA's fate will hinge on the enthusiasm of its political supporters. While that emo-

> tional link appears strong in the White House, it is waning among traditional supporters in Congress, said one veteran congressional staffer. The dispirited reaction of space activists here is that NASA is dying — the question is just how long do we keep it on life support until we pull the plug.[20]

Congressman George Brown, a Democrat from the San Diego area of California, is a long-time supporter of NASA in general and the Space Station in particular. He would rather not see the Station canceled, but if the Clinton Administration does not provide an adequate budget to NASA, he would vote to kill the Station and reapportion the remaining funds to other NASA programs, rather than watch NASA's vitality be sapped away. It is deplorable that the situation has come to this point, because it does not have to be this way. There are two excellent space proposals out there, although they are being ignored by the status quo, because they are not sanctioned by the Beast. Space Industries International has designed, with its own funds, the Industrial Space Facility, which could be built and launched for a fraction of the cost of the Space Station. It was never meant to replace the Beast's projects; it was designed as an interim step to a full space station, so that we could acquire necessary data. Also, it has commercial potential. The second idea is that of forming a space station out of spent shuttle external tanks, which are currently allowed to fall back into the Indian Ocean.

At a press conference in late 1993, I asked Goldin, "With the concentration on cost consciousness at NASA now, why do alternative ideas, such as Space Industries International's Industrial Space Facility and ETCO's and Global Outpost's external tank space station designs, have a hard time getting heard and reviewed and considered at NASA?"

Goldin replied:

> We went through a redesign of the space station. As part of that redesign, we looked at all possibilities. We consulted with the President, we consulted with Congress, and we came up with a redesign. We are not going to redesign the Space Station anymore. We've cut 40% of the life cycle costs out of the Space Station. We now have a program and a schedule that we're marching to. This [sic] has been five votes this year on the Space Station. We are not going to go back to paper; we're going to cut and build hardware. We were so far down the pike that we had a lot of hardware and a lot of designs, and to start with a clean sheet of paper at this point in time might not have been the most expeditious approach. It's not that we don't want to have the most efficient approach, but given where we were, to go back to ground zero would not have utilized the resources we've developed.[21]

Goldin is excellent at the task that has befallen him. He must placate the combined tidal forces of the Clinton Administration, Congress, the NASA bureaucracy, and the American public. He is doing a remarkable job. Yet examine closely the last sentence of his reply. He knows there are better alternatives to the current station design, and although he sidestepped my question, he knew exactly to what I was referring. His goal, however, is to use the already developed resources, not to build an efficient station. The Beast has given him marching orders.

Goldin is like the loyal colonel commanding a battalion that has been ordered to take a certain hill during a battle. He knows that it is not the most effective course of action, and that there will be many casualties. He has a better battle plan, one he knows will win, but like a good soldier, he will follow orders. If he succeeds, his heroic achievement will probably be just a blip in the historical data banks. If he fails, however, the blame will fall squarely on his shoulders. Just like the Challenger Seven, Mr. Goldin is expendable in the greater cause of The Beast.

Perhaps, as the congressional staffers believe, it is time to seriously consider pulling the plug on the old NASA, especially if the Space Station is canceled, and create a new organizational structure which would more effectively move us into space as a nation and as a species. The monolithic structure should be disbanded; the monopoly of NASA should be broken up, as was AT&T a few years ago. In its place should be created space regulatory agencies, modeled on the Federal Aviation Administration and the Federal Communications Commission, allowing private enterprise to actually develop the space infrastructure. Basic research and development should be carried out under the National Science Foundation, or some similar agency. Any activity which overlaps an existing agency's charter should be transferred to that agency. For example, the National Oceanic and Atmospheric Administration should take over the Earth Observing System. Government-industry-academic consortia should be managed under the rubric of the Centers for the Commercial Development of Space (the ones that survived cancellation). Above all, whenever possible, the government should buy data and services from private enterprise, thus stimulating the private development of space. The government should also develop laws and tax policies that encourage investment in the space infrastructure, and otherwise get out of the way. The Omnibus Space Commercialization Bill (H.R. 2731), currently pending before Congress, would, if passed into law, make enormous strides toward creating a favorable climate for commercial space development. I am afraid, however, that the Beast has grown too large and too ravenous, and these are just idle musings.

With each sunset, however, there is the promise of a new dawn. Out of the ashes of destruction and decay rises the fiery Phoenix, to fly into the new day. Gary Hudson, the subject of a later chapter, actually named one of his first space projects *Phoenix*, for this very reason. He is just one of many entrepreneurs and visionaries you'll meet in the following pages. Not content to let the status quo close off the hope of the people, these champions are taking access to space into their own hands. Instead of the monstrous, monolithic "Space Program," which will only ever allow those with "the right stuff" the opportunity to "touch the face of God," they intend to build a network of proactive space enthusiasts and entrepreneurs who can band together into loose coalitions to achieve space projects. We will finally and continuously have the tools at our disposal to build the landscape of our dreams, out there among the stars. Space is finally opening up for everyone. Come along; the Grand Adventure is just beginning.

Notes:

1. Bloomfield, Lincoln P., Ed., *Outer Space: Prospects for Man and Society:* 1962, Prentice-Hall, Inc., Englewood Cliffs, NJ, p. 124.
2. Skinner, Richard M. and William Leavitt, Eds., *Speaking of Space:* 1962, Little, Brown and Co., Boston, p. 210.
3. *Aviation Week and Space Technology,* McGraw-Hill, Inc., New York, November 9, 1992, p. 26.
4. *Space News,* Army Times Publishing Co., Springfield, VA, February 8-14, 1993, p 10.
5. *Space News,* January 10-16, 1994, p. 6.

6. *Space News,* April 18-24, 1994, p. 10.
7. *Space News,* October 11-17, 1993, p. 3.
8. *Space News,* February 28-March 6, 1994, p. 6.
9. *Space News,* October 5-11, 1992, p. 15.
10. *Space News,* January 10-16, 1994, p.4.
11. *Space News,* November 16-22, 1992, p. 27.
12. *Ibid.*
13. *Space News,* July 1-7, 1992, p. 10.
14. *Aviation Week and Space Technology,* March 7, 1994, p. 21.
15. *Ibid.*
16. *Space News,* April 25-May 1, 1994, p. 1.
17. *Space News,* April 25-May 1, 1994, p. 21.
18. *Ibid.*
19. *Space News,* March 28-April 3, 1994, p. 3.
20. *Space News,* April 18-24, 1994, p.1 and p. 21.
21. Space Exploration '93, Houston, TX, Goldin press conference, October 28, 1993.

Two

Why Go To Space At All?

I
The Vision

Space exploration and development is exciting! It is easy to become absorbed in the details, the discoveries, the adventure, and forget why we began such a quest in the first place. If we are ever to reach space as a civilization, it is imperative to understand the minutiae, the nuts and bolts of how it is done. It is, however, no less important to examine why we want to go, what we intend to accomplish, what our hopes and dreams are upon achieving our goals.

I embarked on my quest toward a spacefaring civilization to fulfill a personal vision. I have, from my earliest memories, loved the idea of space. I have always marveled at science fiction, and in eighth-grade science class, where I was racking up a solid C average, I achieved A pluses for the two weeks we concentrated on space.

As I grew into adulthood, other interests absorbed me, and space studies slid to the back burner. Then, in 1983, I participated in a seminar entitled "2013: the world 30 years from now." The task on the first day of the seminar was to envision the state of the world in that future time. On the second day, we had to figure out how to bring it about.

With a fellow attendee I was assigned to go into a darkened room, close my eyes, and relate my vision of the future. Upon shutting my eyes, a fully-articulated vision leapt into my imagination. I saw a re-greened Earth, dedicated to agriculture and environmental parks. There were perhaps six large cities on the whole planet, mainly distribution and collection centers for the solar system's economy. The cities were built downward, into the Earth, with no eye-jarring artificial structures to assault the senses. Ground transportation was achieved by means of magnetic strips between destinations. Vehicles were encoded with their destinations, much like bar-coding, then glided along the strips at tremendous speeds. Since every vehicle was locked onto the strips and traveling at a uniform speed, there were no accidents. Power was provided by clean-burning hydrogen fusion and solar power satellites.

My mind then flew to the Moon. It was one massive industrial park. Every conceivable industry was represented and allowed to prosper in a free and open market. The goods and services produced there were shipped all over the solar system. The Moon's far side was reserved for pure science and astronomy.

I saw great ships plying the pathways of the solar system, visiting the colonies of Mars and beyond, bringing back valuable resources from the nether regions, enriching everyone.

Beautiful pleasure yachts powered by solar sails gracefully wandered about the spatial sea. Huge free-floating space resorts supplied every kind of diversion, from flying under one's own power, to all kinds of space sports, to discreetly appointed zero-g love nests.

The outer worlds of Europa, Titan, and others were being explored for organic matter and even life, while tiny robots of nano-technological origin set about terraforming the worlds where no life previously existed, yet which would be useful to humanity.

A great power generator was in full operation around stately Jupiter, producing energy from Io's interaction with the great planet's magnetosphere. Automated probes with the most advanced hydrogen-scoop and antimatter engines were forging their way to the nearest stellar neighbors to initiate a first reconnaissance. The whole neighborhood of the Sun was bustling and thriving with human activity.

I revived from my vision and eagerly explored ways to bring it about during the rest of the seminar. I did not stop when it was over. Since then, I have not ceased exploring methods of bringing my vision to fruition. With luck and hard work, and banding together with others of similar persuasion, I don't doubt that it will happen.

II
The Economic Argument

The two most prevalent arguments I have heard to support space exploration are that it is humanity's destiny to go to the stars because humans are natural explorers, and that the myriad of spin-off technologies from the space program have significantly improved our lives. While I agree with both of these arguments, they are vaguely unsatisfying as justifications for a multi-billion-dollar enterprise that may span several generations. Just as many spin-offs can be created by investing in high-technology developments on Earth, and if you want exploration and adventure, the ocean trenches await, or you can join the army and be all you can be. It seems the majority of people on the planet are not convinced by those arguments either, leading them to lobby their representatives to address more pressing, earthly concerns, causing space budgets around the globe to dwindle.

The standard counter-argument to developing space is, "How can we justify spending money on space development before we solve our problems on Earth?" Space advocates and opponents throw the various arguments back and forth, and never seem to achieve a resolution. Meanwhile, the space budgets shrink.

There are, however, several compelling arguments for the exploration and development of space. They will be addressed one by one in the following sections. This section is devoted to the most pressing reason to expand into space: the economic argument.

The greatest good a government attempts to achieve for its people is to provide them with a climate in which they may work hard to create a better life for themselves. Although this ideal is a road fraught with many pitfalls, detours, and setbacks, it has been the guiding principle of western democracies for over two centuries now, and with the recent collapse of the Soviet Union, many eastern countries have joined the Grand Experiment.

The best method these countries use to create this climate is an ever-growing economy. We are currently witnessing the damaging effects of stagnant or recessed economies around the globe. People who feel they have lost their opportunities for advancement, or feel others are taking those opportunities from them, are much easier to persuade to hate, to kill, to go to war.

Therefore, many governments consider it imperative that they keep the economy growing, at almost any cost. That cost tends to manifest itself as a fierce battle for market share. We in the U.S. continually complain about our trade deficit with Japan, but I am sure that the Japanese love it.

The opening up of the eastern bloc and the ongoing development of the "third world" make it appear as if there is much more room for growth in the global economy, but ultimately the Earth is a closed system. Life on Earth is fueled by incoming sunlight, and a current theory postulates that the oceans came from water from millions of bombardments of comets, but on the scale of economic activity of humans, the earth is finite in resources. If we try to keep our economy growing forever based on the finite resources of Earth, we will one day run out. The ensuing dark ages will make the last go-round seem like a holiday picnic.

We must keep the economy growing, because the population of the planet is experiencing an exponential increase, and every last one of the new people wants to eat. Most attempts to curb population growth have been unsuccessful, yet it has been discovered that the best method of population control is a high standard of living.[1] And that is achieved through an ever-expanding economy.

The only way to keep the economy expanding infinitely is to expand our resource base infinitely. The universe is a big place. Human ingenuity is such that we will find innumerable ways to economically prosper in space. The list of known methods is already long: solar power satellites, lunar helium-3 production, asteroid mining, hydroponic agriculture, tourism, to name a few.

We only need a few visionaries, such as the ones covered in this book, to realize the magnitude of the carrot of space development in front of them and the stick of global depression behind them, to jump-start the space economy. The explosion of new industries and jobs created in their wake will dwarf any economic expansion that has heretofore occurred in human history. Poverty would diminish worldwide as the growing labor requirements of the new space industries put more and more people to work. Moreover, as we progress into space, new opportunities will be discovered or developed, further compounding the positive economic effects. We will have once and for all escaped the trap of a closed, cyclical economy, and the riches of the solar system will lie before us, like a field ripe with wheat, waiting to be harvested for the benefit of all.

III
The Environmental Argument

One of the most compelling reasons for developing space lies in the necessity of protecting our home planet. Humans are beginning to exert great pressure on the ecosystems of Mother Earth. Even conservative population estimates predict 10 billion people by 2050, twice as many as we have now, with no indication of the growth rate slowing. Over half of the people living now are under 15 years old, and every last one of them is hungry. Food distribution is currently an economic problem, but if we cannot curb population growth, it will soon be a biological problem.

Our industry has developed to a point that we can wield amazing power and accomplish great feats. It all occurs, however, within the earth's biosphere, so any waste products stay right here, creeping into our food chain and atmosphere.

Conservation is a noble cause, but it is ultimately a losing proposition; the best we can hope for is to slow down the rate of pollution and depletion of natural resources. We merely delay the inevitable day of our own destruction. It is impossible to turn the clock back; to get out of our present dilemma, we must march forward.

Science has marched forward, and has devised possible solutions to our problems. Less-polluting energy sources, electric cars, alternative urban designs, to name a few, hold the promise of improving our lives and our chances of survival. Yet we have invested so much in our current way of doing things, both financially and psychically, that our present systems stringently resist change.

The earth is essentially a closed system. Basically everything we know is right here with us. Moreover, we have explored it all, even to the depths of the oceans. The impetus to change from which our ancestors benefited as they discovered new lands and new people is no longer a factor in our lives. The best we can manage in the current state of affairs is an endless redistribution of resources. Our condition is like a candle; it may burn brightly now, but eventually it will run out of fuel, and snuff out.

If one approaches the human condition logically, the path we must follow is obvious. *Question:* How do we provide for the needs of an ever-growing population? *Answer:* Develop an ever-growing resource base. *Question:* Since the Earth is completely explored, where do we find these new resources? *Answer:* Where we have not yet looked: space. *Question:* Space is hard to reach. How will we ever develop such a resource base? *Answer:* As in all previous phases of human development, the pressure of an expanding population and the perception of dwindling resources will drive a few adventurous souls to risk the unknown for the promise of a better life. The rest of civilization will follow in due course.

As we develop a space-based economy, we will be presented with the unrivaled opportunity to start from scratch, to develop systems as efficiently as possible, rather than incrementally imposing solutions on existing infrastructure, thus creating a palimpsest of technologies that is almost impossible to modify. Moreover, these new discoveries and inventions will filter down to Earth, improving everyone's standard of living.

Eventually our space infrastructure will develop to such a degree that we can allocate resources and real estate based on their most efficient use. The Moon, with no ecosystem to damage, can become the seat of heavy industry. The Earth, relieved of its population pressure and industrial burden as people migrate, can be allowed to regreen. The whole planet can be devoted to agriculture and preservation of environment, with only a few strategically located small urban areas to serve as distribution centers. Free-floating space stations can be adapted to whatever purpose the builders have in mind. The benefits of an industrialized society will finally be within everyone's grasp.

As I've already mentioned, the only truly effective method of population control is a rising standard of living. Under such a condition, previously impoverished people no longer feel the need to procreate just to generate offspring to take care of them in later life. The corresponding rise in education and comfort of lifestyle empowers women to take control of their own destinies, and alleviates the pressure on men to impulsively act on their biological imperative to "sow wild oats," which causes the population explosion in the first place. Space development can provide such a climate. Moreover, as space becomes colonized, humanity would finally be a multi-planet species, and an ecological or other type of catastrophe would not eradicate us, as is now the case. Events such as Chernobyl and Exxon Valdez are screaming wake-up calls that we should carefully heed.

There is the counter-argument that humans will take their polluting ways with them wherever they go. This may be true, but if we do *not* develop an off-world economy, we are doomed to drowning in our own filth. Moreover, as we advance into the heavens, we will learn, as we have in our past explorations, to treat our environment and our fellow humans with an increasing degree of respect and care. Human development is not a linear process. There have been tragedies and travesties, such as the extermination of most of the indigenous peoples as Europeans advanced into the "new world." No evil was perceived at the time, but we have now learned that grave lesson, and we are much less likely to inflict similar damage in the future.

One cannot advance into space without considering how to eat, how to excrete, how to breathe; in short, what it means to be *alive*. And one cannot examine those aspects of living without gaining a new appreciation for life. The advance into space will make us more ecologically aware, just as the photo from Apollo 8 of the fragile Earth floating in a sea of black became an icon for the environmental movement. Space *is* our environment. Our molecules originated in the stars, and now our bodies, minds, and spirits must return to space, to the source of our existence. Only then will we truly be able to understand and care for our beautiful, precious Earth.

IV
The Education Argument

I have a brother-in-law named David who teaches middle-school English in a rough neighborhood on the south side of Dallas. One day he was hammering away at prepositions, and the kids were just not getting it. Suddenly he had a brilliant idea. "Where does a cockroach go?" he asked. Responses flew back. "Under the refrigerator!" "Through the food!" "In the wall!" "Around any bug spray we put down!" David responded,

"Exactly. A preposition is anywhere a cockroach can go: under, through, in, around."

This is a striking, if a little morbid, example of what is needed to reinvigorate education. David taught those kids something, because he made it relevant to their lives. An education devoid of meaning will continue to crank out people who are functionally illiterate, and who don't possess the basic skills to get and hold even manufacturing jobs, because they have no *reason* to learn.

I was the same way in junior high school, at least in relation to science. I was getting C's and D's in my science class, until the two-week section on space showed up. For those two weeks, I got straight A's, then went right back to C's and D's when it was over. My science teacher asked me, "What's the deal? Why can't you keep your grades up all the time?" "I guess I'm just into space," I replied.

I am not the only kid who was or is into space. Witness the popularity of *Star Trek*, *Star Wars*, and science fiction in general. The audiences for these books and movies contain a large number of kids who think space is "cool." Yet the educational institutions, by and large, are not tapping into that natural interest.

There are a few notable exceptions. The Challenger Center is an outstanding example of the use of space as an educational tool. Now consisting of eighteen centers around the country, with an average of six added each year, Challenger Centers simulate the experience of space missions for more than 200,000 students a year. Students must cooperatively engage in various tasks and solve problems aboard a model of a space station, or from the mission control. One participant remarked, "Before I came to the science center, I didn't really like science, but when I got to the science center it changed my mind. I really want to come back again."[2]

Another popular space learning tool is Space Camp at the U.S. Space and Rocket Center, next to Marshall Space Flight Center in Huntsville, Alabama. Thousands of students each year engage in simulated astronaut training, and come away with a huge enthusiasm for space. Many have gone on to receive advanced degrees in space-related fields, and some have even become astronauts.

Kids love space. Space is cool. Space is *real*, as real as the kids' dreams to fly in space, to go to the Moon, or even to Mars. Kids are motivated to learn math and science if they can see a definite, real purpose for their education. More programs like the ones cited above should be instituted.

These days the most practical application of space education is to stimulate math and science learning, but there is no reason that the space environment cannot reinvigorate all disciplines. Students can read and write about space, they can study the history of the world and how it led to space exploration, and can practice logic and philosophy in examining where this Grand Adventure will ultimately lead us. They will have a *reason* to learn, instead of collecting esoteric and unrelated facts. If we do it right, then some day these children, in whom we instill a love of the stars, will take us to them.

V
The Political Argument

From the beginning, space has been a political tool. President Kennedy goaded us to the Moon on a dare to beat the Russians. And space continues to be politicized. Except for a robust communications satellite market, space is primarily the province of governments. Based on a left-over cold war paradigm, the U.S. government uses its influence to discourage private development of space. NASA has given little to no support for concepts such as

Space Industries International's Industrial Space Facility, Gary Hudson's commercial spaceplane, and various private, external tank space station designs. Only nominal support was given for the Commercial Experimental Transporter (COMET), which was recently canceled, and a recent NASA report advocated abandoning the Spacehab commercial shuttle middeck experiment platform.

A government is a conservative beast, always looking to protect its position and guard against any possible aggressors. It is therefore not surprising that the U.S. government sought to limit and control space access during the cold war. The historians can debate whether that was a prudent course of action, but the cold war is now over. By adopting an open, rather than restrictive, policy toward space development, the U.S. government, and other governments around the world, can now *increase* their security and ability to provide for their citizens.

Modern political activity has three main goals: 1) to defend the territory of a country and its citizens, and to establish, if possible, a peaceful environment, 2) to create and maintain an economic framework in which the people can prosper, and 3) to provide leadership and a sense of direction and purpose. Space can provide all three of these elements, in a sustainable and non-aggressive fashion.

One of the prime causes of war is the perception on the part of a government or people of a lack of a certain desirable element, such as food or freedom, and the only way to get it is to take it away from some other entity by force. We have entered the Age of Claustrophobia, in which all the surface of the earth has been explored, and population densities are rising, so such wars will only increase. Modern wars of attrition can damage the winner as much as the loser, and the experience is psychologically draining. Consider the still-extant official policy of the U.S. toward the recently-defunct Soviet Union of Mutually-Assured Destruction, or MAD. Such a policy can ruin an economy.

A thriving space economy would remove any need for such ludicrous policies. A country with colonies and enterprises in space could not possibly be totally destroyed. There would always be someone left to retaliate. More to the point, however, a country that is deriving a large share, if not most, of its wealth from the broad expanse of space, where there are no borders to protect and plenty of room for expansion, will feel much less compelled to engage in conflict rather than just move on to the next asteroid. With an ever-expanding space economy, the only obstacle to obtaining desired elements will be the industry of the people involved. Finally, a government which can motivate its people to seek their fortunes among the planets and stars will not only benefit from the expanded tax revenue base, but will be providing a positive goal for them. It is a psychological axiom that a positive motivator is more effective than a negative one. Therefore, a government sponsoring a focused program of economic expansion into the heavens will unlock a creative tidal wave in its people.

It is also a psychological axiom that a negative motivator is more effective than none at all, and that is the current method employed by most governments. Despite Kennedy's high-flying rhetoric, the motivation behind Apollo was to beat the Russians, so they would not have an advantage over us. The American people were motivated to great achievement, but twenty-five years later, we have little to show for our national endeavor. We have no space infrastructure, NASA is aimlessly wandering into the future with mounting failures, and the American public has even soured on space ventures as a whole.

The Grand Debate of the twentieth century concerned itself with answering the question, “Which system is more effective, capitalism or communism?" Capitalism has won that debate. We must now take the lessons from that struggle and apply them to space development. As it stands now, space development in the U.S. and all other spacefaring nations is a monopolistic, government-run affair. Not coincidentally, all spacefaring nations are scaling back their space projects. All nations must learn the lesson that there is great potential lying dormant in their citizenry, which can be unleashed if the people are led to perceive that it is to their benefit to develop space. That benefit can be most directly expressed as profit. Give people the positive motivator of profit, and you will not be able to stop space development. You also will not be able to stop robust economies and rising standards of living and tax revenue bases, high expectations and positive feelings about the future, expanded international cooperation and decreasing desire to engage in warfare, and decreasing dependency on governments for well-being and welfare. To some governments this situation is anathema, for they perceive that they will lose power and prestige, but those reactionary institutions will fade away, and the nations who encourage such activities will prosper.

We are truly at a position in history when we can fundamentally change the way governments interact with their citizens and with each other. Forward-thinking leaders will shift their paradigms and release the potential for unlimited economic growth latent within the people. History is not a smooth course, however, and there will be reactionary bumps along the way. Some leaders will find it difficult, if not impossible, to curb their power-hungry tendencies. Nevertheless, as one of my favorite history teachers once told me, "There is nothing as powerful as an idea whose time has come." The time for the stars has come. Our leaders must open up the road to the future, or they will be unceremoniously dumped into history's potholes.

VI
The Historical Argument

The flow of the river of history is not as smooth and steady as it may appear at first glance. There are floods and droughts. The river's channel may change abruptly, transforming a promising waterway into a stagnant bayou. There are deep currents, eddies, blockages, rapids, and waterfalls. Yet the inevitability of the flow of history can cut grooves in the human psyche as deep as the Grand Canyon, and eventually the waters of time reach the cosmic ocean. It is best to learn the river's course, and to flow with it.

In 1001 A.D., driven by population pressures, Viking settlers reached the shores of what is now known as Newfoundland, off the east coast of Canada, which they called Vinland. After less than a generation of colonization, they were driven away by skirmishes with "Skraelings," or native inhabitants, and never returned. Word of their discovery never reached beyond the boundaries of Norse culture, because they considered the find insignificant. They had no grand plan of discovery and exploitation; they were merely searching for "living space," and when they found a land already occupied, they turned around. It never occurred to them that they had discovered some place "new;" the land was very similar to that with which they were familiar. Why fight the ugly Skraelings for a patch of land not much better than the one they had left? Thus was concluded one of the greatest dead-ends in history. The river changed course, and the Norse dropped from the history of discovery.

In the early fifteenth century, China under the Ming dynasty engaged in a systematic program of maritime excursions. The purpose was to demonstrate to the rest of the civilized world (at that time mostly Muslim) that China was the center of the earth, the only true civilization, and that they needed nothing from any other culture. In fact, in their interactions with "the barbarians," they gave more "tribute" than they received, ostentatiously revealing the bounty and wealth of "the Middle Kingdom." Huge, opulent ships sailed the coasts of present day Southeast Asia, India, Arabia, and all the way to the east coast of Africa. The whole world knew of the grandeur of China.

In 1424 A.D. a new emperor was crowned, one who focused internally, and did not perceive the value of the sailing expeditions. He concentrated on consolidating his power at home, and in 1433, issued an edict recalling the great fleets and forbidding the Chinese to travel abroad. The historian Daniel Boorstin tells us, "The bureaucrats sensibly argued that the imperial treasure be spent on water-conservation projects to help farmers, on granary projects to forestall famine, or on canals to improve internal communication, and not on pompous and reckless maritime adventures."[3] The Chinese trapped themselves in an eddy of the river of history and, as Boorstin concludes, "Fully equipped with the technology, the intelligence, and the national resources to become discoverers, the Chinese doomed themselves to be the discovered."[4]

Recovering from the long twilight of the dark ages, and with gold glittering in their eyes from their experiences in the Crusades and from the tales of Marco Polo, the Portuguese embarked on a systematic program of discovery with one goal: to profit greatly from bringing the riches of the Orient to the West. The effort was spearheaded by one visionary man, Prince Henry the Navigator. He developed and carried out a systematic series of expeditions to travel further down the African coast than any Europeans before then. There was psychological resistance from his captains and seamen, born of fear and the weight of tradition, but Prince Henry's rewards were so great that the power of gold overcame the fear of the unknown.

When Prince Henry died, leaving a legacy of great discoveries, and more importantly the idea that discoveries *could* be made, the torch was picked up by Fernão Gomes. He negotiated a contract with the King of Portugal to make discoveries further down the coast of Africa, in exchange for a monopoly of the trade he generated. With the profit motive pushing him, Gomes discovered as much coast in five years as Prince Henry had discovered in thirty. The Portuguese actively pursued the knowledge and benefits of discovery; they sailed with the flow of the river of history.

The ultimate profiteer was Columbus. The Portuguese rounded the southern tip of Africa in 1488, and Columbus took that as his cue. He finally convinced Queen Isabella to finance his voyage by portraying the spectre of a Portuguese-dominated sea trade to her. After fourteen years of struggling to convince the crowned heads of Europe of the profitability of his venture, he finally got his chance, and made history's biggest and most important mistake in 1492. He died not realizing he had not achieved his goal of reaching the Orient. Nevertheless, he profited from his discoveries, as did all of Europe. Even though his facts were off, Columbus knew that just around the next bend in the river of history were glorious and enriching discoveries, waiting for someone to sail out and grab them.

The lessons to be learned in our own era of space-age discoveries are clear. We should not, like the Vikings, haphazardly wander into the void; it may beat us back, and offer no reward. Apollo was such a dead-end. Apollo and Vinland were both politically-motivated expeditions, and both died when the political issues faded.

We should not engage in exploration or development for its own sake, or to show off to the rest of the world, like the Chinese.

We already hear clamors from Congress to kill the space station and "spend the money at home," just like the Ming bureaucrats.

Economic return is ultimately the only valid reason for space exploration and development. Everything else — science, environmental monitoring, technological advances — can and will be spun off of the drive to find profit among the planets. The discoveries of Columbus and the other European explorers fueled the greatest advances in science and civilization, advances which have not yet ceased improving our lives. True, many evils were perpetrated in the over-zealous spirit of discovery, and whole civilizations were wiped out in the process, but there are two main differences between that era of discovery and this one: first, there are no lives or civilizations to destroy in our solar system (the question of life on Mars has not been totally resolved; I therefore advocate we learn as much as possible about the Martian geography and ecosystem before we engage in any grand plans of terraforming). Secondly, we have the benefit of advanced historical knowledge and appreciation to guide us. We know more about our world and our past than our great exploring forefathers, thanks in part, already, to the increased communications provided by the economic exploitation of space, in the form of communications satellites.

Let us use the full power of the information at our disposal, and let us analyze and evaluate the lessons of the past, as we sail down the river of history into the future. Let us not shrink from making the tough value judgments we must make, if we are to succeed where others failed. Let us not fear the future; let us rest secure in the knowledge that the flow of the waters of time will reach the cosmic ocean, and, equipped with the strong rudder of the wisdom of history, we will sail with the currents into the sea of space.

VII
The Frontier Argument

For all practical purposes, the Earth has been explored. To be sure, there remain facets of our planet that are undiscovered or not fully understood, but we now know the boundaries of our world. No grand, paradigm-shattering discoveries remain to be made, such as those of Marco Polo or Columbus. We have no more frontiers on this planet.

Humans have carved up the Earth like an orange rind. Maps no longer contain "terra incognita;" men (yes, it is mostly men) have drawn political boundaries over the whole globe. There are no unclaimed lands for humans to expand into. Therefore, if one country feels it needs more space or more resources, it *must* take them from a neighbor. War has become a zero-sum game (as if it were ever anything different).

We have entered the Age of Claustrophobia. The symptoms are apparent all over the world: Serbia, South Africa, Rwanda, Somalia, Cambodia, Tibet, Russia, Quebec, Korea, Mexico, Northern Ireland. In fact, almost no one is immune. Deprived of frontiers, we have become bored, and scared of the future, and we take out our frustrations on each other.

In his 1992 book *The End of History and the Last Man*, Francis Fukuyama postulates that, with the fall of the Berlin Wall, history ended. His thesis holds that all the great challenges of history have been played out, and that our greatest challenge in the future is to manage our planet, resources, and people with ever finer degrees of precision. I fundamentally reject this analysis. Such thinking can only contribute to the Age of Claustrophobia. His book is yet another in a series of books that speculate on the future without factoring in space exploration. These books, which include Paul Ehrlich's *The*

Population Bomb, Jonathan Schell's *The Fate of the Earth,* and E.F. Schumacher's *Small Is Beautiful,* take a fundamentally pessimistic view of humanity's ability to solve its problems. The only thing that will save us, they generally posit, is central planning and an endless redistribution of resources. Such a steady-state system, however, would have to be run at 100% efficiency; if any mistakes or disasters occurred, the whole global system would necessarily lose some of its capability and resilience. The entire planet would ultimately run down; the second law of thermodynamics allows for no other result.

We must infuse our world with hope for a better tomorrow, for a chance at improvement and adventure. We must keep as many options as possible open, for no one can predict the future with certainty. In the past, frontiers have provided such hope and incentive. The westward expansion of America has stimulated more hope and more change than anything else in recent history. Granted, many peoples were trampled and destroyed in the rush to fulfill "manifest destiny." A new expansion into the frontier of space, however, does not possess that baggage. Moreover, it has the potential to help correct the evils of the last frontier. The economic explosion generated by the expansion into space will provide innumerable jobs, and eventually many people and industries will live off-world, easing the pressure on Earth. Those disenfranchised peoples who wish to return to "the old ways" will finally have the breathing space to do so.

Consider the alternative. If we do not move into the space frontier, our wars of claustrophobia will become increasingly worse. We are currently locked into the zero-sum game of the boxing match, in which the only way to victory is to beat one's opponent into a bloody pulp. We must replace that way of thinking with the marathon paradigm, in which one is still in competition with others, but the only way to win is to improve one's own excellence. Even though only one participant crosses the finish line first, each player wins, for having stretched her or his personal potential.

A frontier is a good substitute for war. Human beings, especially young ones, are loaded with energy that they just want to use, somehow. The older generations have been very skillful in channeling that energy into aggression against neighboring peoples. The challenges of the space frontier, however, are even more demanding and rewarding than winning a war, because not only does it require energy, but also creativity. Who knows what marvels humans will produce as they advance toward the stars? Finally, by opening the space frontier, all of our cultural "eggs" will no longer be in one basket, minimizing the damage of catastrophes that would ultimately destroy a steady-state world.

If we do not open up the space frontier, we are doomed to ever-dwindling resources and increasingly destructive wars. If we go to space, as Kennedy told us, we are not guaranteed success, but we have a running shot at it, and we are destined for failure if we do not. To paraphrase Ayn Rand in *Atlas Shrugged*, it is the finest quality of humans to be able to act. To not do so, when the path is clear, or to be paralyzed by fear or doubt, is the ultimate blasphemy against human nature. The space frontier lies open before us; we have only to hitch up our wagons and head out on the trail to the stars.

VII
The Cosmic Argument

The universe is big — mind-bogglingly big. We have a mind, however, miniscule in comparison to the cosmos, that *can* be boggled at the size of the universe. A frequent comment about the universe runs, "When you consider the size of the stars and the galaxies and the universe, we humans seem to be as puny as amoeba. What right do we think we have to move out into the cosmos?" Those amoeba, however, to which our generic observer compares himself, are busy searching for food, and do not bother with such questions. An amoeba has absolutely no conception of rights, much less whether it has any in relation to the universe. It merely acts according to the promptings of its biochemistry.

The universe diminishes in size as far beyond the amoeba as it increases beyond us. Why, when we contemplate the infinitesimal aspect of the universe, do we not feel correspondingly omnipotent? The answer is, of course, simple: we can no more comprehend the minuteness of the universe than we can grasp the vastness of it. It is outside our scale of reference. Similarly, the speed of light and absolute zero set boundaries we cannot mentally cross, as does the Big Bang and whatever its antithesis is (we do not even know if there is an antithesis), as does birth and death. We have been given an arena in which to act, just like our diminutive amoeba friends.

The world's religions have addressed themselves for millennia to these issues. They all claim to be correct interpretations of the universe, yet they all have to deny certain facts we have discovered through our observational and experimental abilities, in order to construct their internally cohesive, yet often flawed and inconsistent, portrayals of the cosmos. Ultimately, despite all our "advances," we are no closer to understanding the true nature of the cosmos than the blissful, innocent, ignorant amoeba.

Yet we walking apes, with one finger very usefully turned against the others and a three-pound computer in our heads, have figured out how to crudely manipulate aspects of our universe. We have somehow developed the amazing capacity to ask "why?" and "what if?" It is a marvelous capacity to be able to question the universe. This capacity of curiosity has created more sublime beauty and more horrible destruction than any other aspect of our nature. We do not know the source of it, but it is there, and it is ours to use.

The question, "What right do we have to advance into space?" is really asking, "How can we be sure that this capacity of curiosity, the source of which we are unaware, will not destroy us, once we have left familiar surroundings?" It is ultimately a question of fear. The converse to that question, the question of hope, is, "What awaits us out there?" We could very well perish. We could very well thrive. We will never know, however, until we go. We must answer the question ourselves.

The universe, as a material entity, does not care either way; it has not even the capacity to care. Despite all arguments to the contrary, the question, "Does God exist?" has not been definitively answered, yea or nay. If there is a God, he/she/it has not put any limits on our ability to get to space that we are aware of. We will never know, however, until we go. We must answer the question ourselves.

The question of fundamental rights, and of morality, is really asking, "What will add to my life and my chances of survival, and what will detract from it?" The universe has provided us no rule-book. We must answer the question ourselves.

If we do find fundamental limits out there, we will be beaten back. Some of us will perish in space, even if we do not encounter any limitations. That is the nature of experimentation. It is the same trial-and-error method which taught our ancestors which berries to eat, and which to avoid. We will never know until we go. We must answer the question ourselves.

Why go to space? Sir Edmund Hilary would answer, "Because it is there." To that simple yet profound reply, I would add, "Because we are here."

IX
The New Mythology

Ages ago, our ancestors gazed up at the night sky, and saw a panorama of gods. There were no city lights to dim the awesome impact of a million shining suns, the patterns of which arranged themselves into a veritable pantheon. The gods were much closer then, populating the heavens, or lurking just beyond the campfire, or writhing below the waves, or carousing through the forest. Humanity was at the mercy of these capricious deities, so one had better step lightly, and offer regular prayers and sacrifices to the forces of nature, in order to appease the masters of fire, wind, rain, and earth. The various mythologies that arose were depictions of the battles and schemes of the gods, representative of humanity's titanic struggle to survive in an unpredictable world.

Modern civilization has, to a degree, won that struggle. We can now peer to the edge of the universe and of time. We can gaze inside the tiniest particle, and make it dance to our whims, to release incredible power. We have illuminated the night sky, almost blotting out the stars which formerly guided our lives. But we still have not been able to illuminate the deepest recesses of our minds, where the gods live. And we have not succeeded in piercing the thin yet impenetrable barrier of death, that "undiscover'd country, from whose bourn no traveller returns,"[5] as Hamlet mused. The gods are very much alive there, just beyond the boundary of consciousness, yet they take different forms, fight different battles, rule different realms.

They nevertheless hold to an eternal purpose: to shed light upon the nature and fate of humanity.

The gods and their consorts are no longer lords of the air, ocean, and underworld, no longer satyrs and nymphs, gorgons and cyclopes; they are aliens of a million different forms, inhabiting a billion different worlds, as alien to us as the gods ever were. They may be omnipotent, or destructive, or repulsive, or beautiful, or so completely different from us that we have no reference by which to understand them. In short, science fiction has revived the gods, out there, amongst the stars, which science killed here, on Earth.

Joseph Campbell, the noted mythologist, postulated four functions of a mythology. "The first function of a mythology is to waken and maintain in the individual a sense of wonder and participation in the mystery of this finally inscrutable universe.... The second function... is to fill every particle and quarter of the current cosmological image with its measure of this mystical import.... A third function... is the sociological one of validating and maintaining whatever moral system and manner of life-customs may be peculiar to the local culture.... A fourth... is the pedagogical one of conducting individuals in harmony through the passages of human life, from the stage of dependency in childhood to the responsibilities of maturity, and on to old age and the ultimate passage of the dark gate."[6] He could very well have been describing the themes of great science fiction, such as *2001: A Space Odyssey* or *Stranger in a Strange Land.*

One of the classic *Star Trek* episodes, "Who Mourns for Adonis?," captures the essence of this phenomenon. The Enterprise crew are captured by an immensely powerful being and coerced to beam down to a planet. Their captor appears as Apollo, who is an alien that was once revered as a god on ancient Earth. Kirk and the crew can no longer worship the being, but are no less impressed and awed by his power. Defeated by the power that the humans have themselves acquired, Apollo laments to his fellow gods/aliens that there is no more room in the human heart for gods.

Yet the Enterprise sails the shores of the cosmic ocean, encountering the same type of beings that Odysseus did several millennia earlier. The gods have merely changed their attire. The boundaries of our universe have expanded from just beyond the campfire to the edges of space and time, yet beyond the fringe is still an area where the gods may roam freely. As we wander out into the cosmic night, first mentally, through the literature of science fiction, then physically, in the ships we will one day build, we will have to confront the mysteries of life and the unknown, and of our own souls, mysteries no less fearsome, awe-inspiring, and wonderful than when we first looked up at the stars, and witnessed the dance of the gods.

X
The Fear

The land burns under the scorching fires of atomic radiation. There are not many crops and plants left to burn; they died out many years ago, in the eco-catastrophe of a runaway greenhouse effect, and under the searing ultraviolet rays that broke through the fragile, weakened ozone layer. That travesty took out most of the humans and animals, too. The coastal cities had long since been flooded by melting polar icecaps. The few people left banded together into neo-tribes, and proceeded to beat the snot out of one another. A few still had access to "the high magic," and set off some of the nuclear bombs that still actually worked. The vultures, slugs, and cockroaches were having a field day.

This scenario, or a similar one, has been re-enacted countless times on celluloid. Go down to your local video store; it is full of them. *Terminator*, *Planet of the Apes*, *Mad Max*, *Road Warrior*, *Soylent Green*, *The Stand*. The list goes on and on.

The human mind falters when contemplating the future. It is incredibly difficult to effectively plan tomorrow, much less imagine a positive future in which the human race thrives. Our modern media do not help much, either; they like to report sensational events. Thank God death and destruction are still the exception, and not the rule. Still, if one were to believe the endless video feed, the world is destroying itself.

The future can be fearsome, because it is the unknown. We know with a high degree of certainty that we are going to die (the science of statistics teaches us we cannot be absolutely certain of anything; nevertheless, I don't think Lazarus Long is about to raise his hand), but we do not know when, or where, or how. Our daily race is an attempt to answer the question, "Can I find happiness, and can I provide for my offspring, before my cosmic lottery number is drawn?" In this seemingly fickle universe, the answer is often, "No."

A future of pain and destruction is fearsome, yet easier to imagine than a future of happiness and plenty. A positive future requires personal responsibility and action, both of which are foes of that principle of the second law of thermodynamics called entropy. The concept of entropy holds that everything in the universe will attempt to achieve the minimal amount of energy usage, therefore the minimal amount of organization. The existence of humans is a direct affront to entropy, and it won't let us forget it.

Entropy is trying to prevent us from reaching the stars. If it succeeds, we will be stuck here on Earth forever. The Age of Claustrophobia will become the ruling principle of interaction between humans, and we will beat on each other until there is no one left to beat on, or there is no one left to do the beating. We will not be able to maintain any kind of material, much less social, organization, because we will spend all of our time scrabbling for dwindling resources. Our neighbors will be trying to take our meager goods away, instead of looking for or creating their own, because we will all be bearing the yoke of the zero-sum game mentality.

Eventually, the human race will wind down into decay, or will explode in a great conflagration, and will perish. We will have lost the great challenge we accepted when we first asked, "Why?" And the universe will not care.

XI
The Hope

Someday, a father may take his daughter outdoors into a starry night, point up to the sky, and say, "See that star, honey, right next to the really bright one, there? That is the star that we came from, long ago."

"What do you mean, Dad? We've always lived here."

"No, no, I mean people. Many centuries ago, people left that star, searching for new worlds to live on. This was one of the worlds they found. You know that picture in the hallway of your great, great grandfather? He was the first person to land on this world. He made it our home."

"Can I go back to that star someday, Dad?"

"Sure, honey, and you can go anywhere else you'd like. The universe is wide open to you."

The history of humanity is not over; it has barely begun. We are living at the fulcrum of time; the decisions we make in our lifetimes will determine the course of many generations to come. It is a great opportunity to be able to influence the course of the universe, yet it is just as immediate and personal as the question, "How will I feed my children tomorrow?"

Space is our playground; it is our home. Just as life on land began in and was nurtured by the sea, so all of Earth began in and was nurtured by the Cosmic Ocean. Venturing outward will be the greatest challenge our species will ever face, yet we will simply be returning home.

Many people have declared that space is our destiny. That is true, but it is not the whole answer. Destiny implies challenge, and struggle, and hardship. To be sure, those are before us, but we are not adversaries of the cosmos. Space is where we *belong*. We *are* space, and space *is* us. We would not exist had not some star gone supernova, somewhere, sometime, and created the shockwave which set the particles in motion that finally coalesced into our solar system. Then life originated on Earth, and eventually evolved into us. We are the eyes, the ears, the *mind* of the universe; we belong here. Going to space is merely following our instincts.

We will advance into space, and we will survive and thrive as a species. Why? Because we *want* to. For over forty years, the United States and the Soviet Union had enough missiles aimed at each other to blow the world up 25,000 times over. We had early warning systems, extensive intelligence organizations, and an extreme distrust of one another. Yet we are here today to talk about it. We opted for the marathon of space, rather than the boxing match of nuclear war. Our instincts for survival guided our minds, and we prevailed.

Such will be the case with the new frontier. We will perceive the magnitude and the stakes of the challenge before us. We will realize that, to survive, it is imperative to grow, to embark on the unending marathon of expansion throughout the Cosmos, rather than becoming trapped in the futile boxing match over finite terrestrial resources. Once again, our instincts for survival will guide our minds, and we will prevail.

Grave dangers beset our world, most of them of our own design. The Fear is pursuing us relentlessly, and we must never forget that we *can* fail. It is, however, by no means inevitable. We have the chance, right now, to open the solar system, then the galaxy, then the universe, to our progeny. The glories, the beauty, the wonder that await us dwarf anything we have heretofore encountered.

It is not our right to go to the stars; there are no rights that we do not grant ourselves. It is not our moral imperative to expand into space; morality is just the refined clothing of survival instincts, and we can die as easily as we can prosper. Rather, it is our *nature* to move outward, for it is nature which brought us into being on Earth, through the intricate workings of the entire universe. Through the centuries, we have slowly gained a more accurate perception of the universe, and now, for the first time, in *our* lifetimes, we have the means to leap off this planet and explore the heavens. The door is open; it is time to go home.

Notes:

1. *Business Week,* McGraw-Hill, Inc., NY, May 23, 1994, p. 16.
2. *Ad Astra,* National Space Society, Washington, DC, September/October, 1993, p. 27.

3. Boorstin, Daniel J., *The Discoverers:* 1983, Vintage Books, NY, p. 199.
4. *The Discoverers,* p. 201.
5. *The Globe Illustrated Shakespeare:* 1983, Crown Publishers, NY, p. 1885.
6. Campbell, Joseph, *Historical Atlas of World Mythology,* Vol. I, Part 1: Harper and Row, NY, pp. 8-9.

Part II
The Solutions

Three

International Space Enterprises

The next time humans land an object on the Moon, it will not be under the auspices of the United States Government. International Space Enterprises (ISE) of San Diego, California, plans to land a privately-funded craft on the Moon by 1997. ISE's mission is "to develop and conduct commercial international space science and exploration projects which will reduce costs, stimulate business investment, and improve U.S. international relationships."[1] The company intends to be a full-service lunar transportation and mission support company, and hopes to reignite the spirit of lunar exploration with its activities, recalling the heady days of the Apollo landings.

To achieve these lofty goals, ISE has teamed with Lavochkin Associates of Moscow, the designers of the Russian Lunakhod rovers and the probes sent to the Martian moon Phobos in 1989 (both probes were subsequently lost during operations). The two organizations have formed a joint stock company called ISELA, which will share the profits of the lunar venture. The joint company will use Russian technology and hardware to develop a fleet of lunar landing vehicles. The first will be the ISELA-600 lander, which will be capable of placing 600 kilograms of equipment on the lunar surface. Later, a larger ISELA-1500 will be put into service, which will be capable of landing 1500 kilograms on the Moon.

Both craft will support the Universal Payload Adapter, which will accommodate up to fifteen payloads per launch. Such an approach, combined with using relatively low-cost Russian Proton launches and the ISELA landing vehicles based on existing Russian hardware, will make access to the Moon eminently affordable for the government organizations, research institutions, schools and universities, and private companies who are anticipated as the market base. ISE Vice President Larry Bell predicts, "We can do it for a tenth of the cost of NASA doing it."[2] Work on the landers is well underway; on March 18, 1994, the company unveiled a one-tenth scale model of the ISELA-600 lander, developed by its Lavochkin partners from existing hardware and technology.

The eight planned missions will be launched, on average, twice a year, carrying such diverse equipment as telescopes, rovers, stereovision cameras, sample return and analysis devices, geological instruments, and exploratory spacecraft. Scientists and researchers who have been waiting decades for a return to the Moon will finally be able to get their experiments to lunar space and to the surface of the Moon. Through Lavochkin, the joint company has gained access to the Russian Proton launch vehicle for delivery to the Moon. ISE has also enlisted the cooperation of Rockwell International Corporation, builder of the Space Shuttles, to aid in designing and planning the project. Using off-the-shelf technologies provided by its partners, ISE will be able to offer its lunar delivery service for a fraction of the cost NASA would charge for comparable operations. Customers can design their own payloads, or save more money by contracting with ISE to provide data from their own sources. ISE will also market scientific data generated by company-owned and operated instruments.

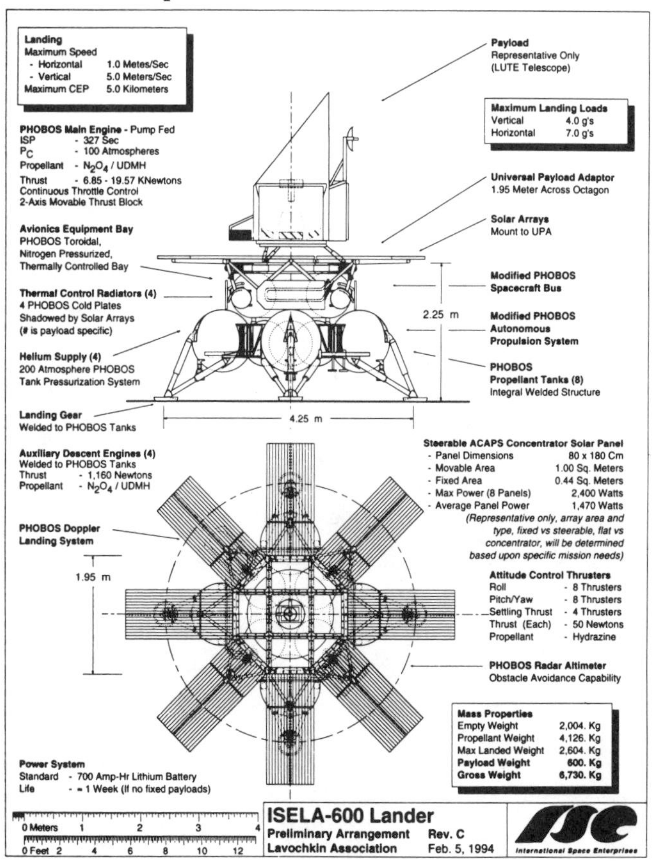

Figure 3-1

The ISELA-600 Lander is expected to land on the Moon in 1997. (Illustration courtesy of International Space Enterprises.)

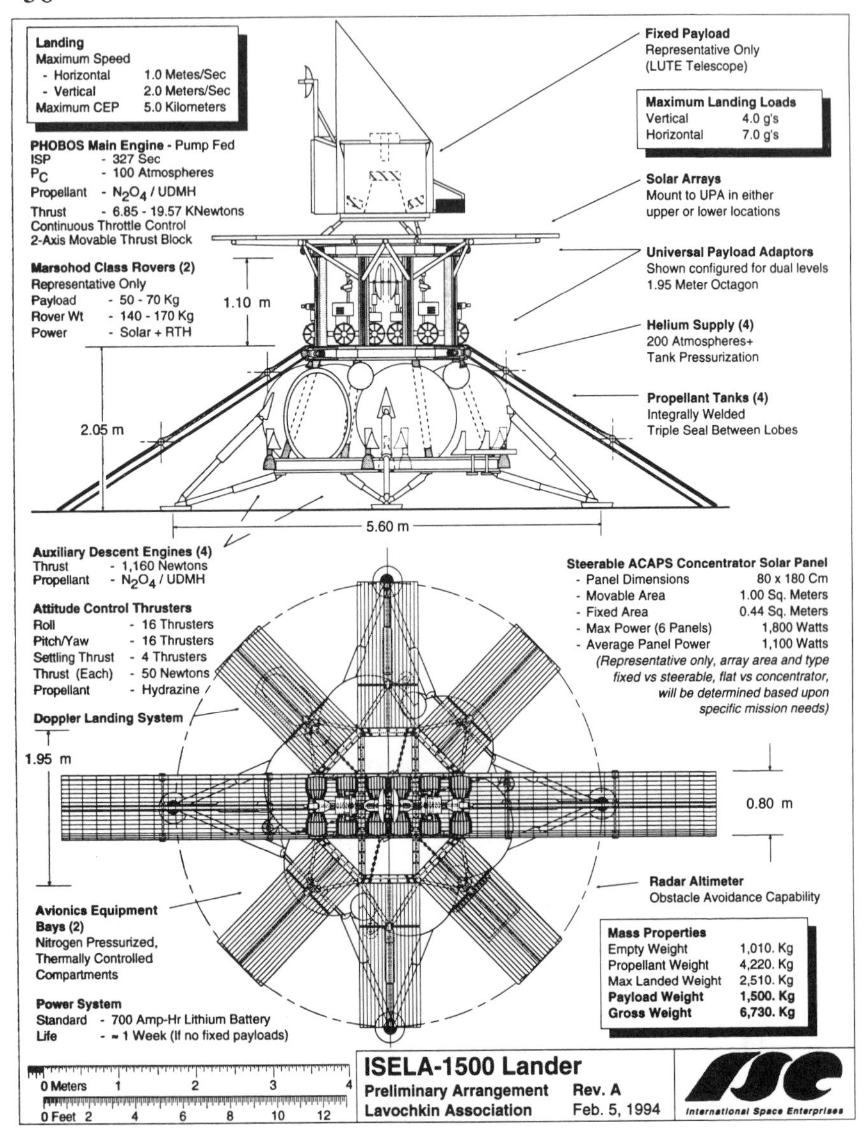

Figure 3-2

The ISELA-1500 Lander is being developed through a private joint-venture between a U.S. company and a Russian company. (Illustration courtesy of International Space Enterprises.)

"We are here today to announce that preparations are underway for our first mission in July of 1996 and tickets are going on sale," President Michael Simon announced at the company's first press conference on September 29, 1993.

> Our unique combination of reliable and cost-effective international space transportation systems will enable us to generate the economies of scale necessary to make space exploration truly affordable. There has been significant interest from university, government and commercial users alike. Our program takes advantage of a new spirit of U.S.-Russian cooperation to open up new frontiers of lunar science, exploration, and business. It will promote private sector investment in science and technology programs vital to U.S. competitiveness; will channel capital into high-wage, high-skill jobs; will produce technology spin-offs that will create new science and business markets; and will reinvigorate our children's interest in science and technology by enabling students at all levels to actively participate in lunar and astronomical science projects. These same actions, in addition to benefiting the U.S., will aid in the stabilization of the Russian economy while ensuring space assets are used for peaceful scientific purposes.[3]

The company boasts an excellent pedigree. Founded in 1992, it is the brainchild of several former General Dynamics employees. Michael Simon, formerly a manager of advanced projects at General Dynamics, is President and Chairman of the Board. David Mazaika, who was involved with international business development and strategic planning at General Dynamics, is Vice President of Business Development. Tom

Kessler, a project manager at General Dynamics, is Director of Mission Integration. These visionaries are joined on ISE's board of directors by Larry Bell, who is Director of the Sasakawa International Center for Space Architecture at the University of Houston; Nathan Goldman, adjunct professor of space law at the University of Houston, who has written several books on commercial space development; and Valery Aksamentov, former Senior Research Fellow at the Moscow Aviation Institute. With their combined experience and connections throughout the aerospace industry, these gentlemen hope to challenge the status quo of that industry, and open up the era of private commercial space business.

ISE's first mission is scheduled for July 1996, when the company will deliver various scientific instruments and a communications satellite into lunar orbit to facilitate transmission of data and communications from subsequent missions. In 1997, the company plans to accomplish the first lunar landing since 1972.

The anchor tenant for that mission is a company called LunaCorp, which intends to land a rover that will, among other things, drive by the site of the Apollo 11 touchdown (see the next chapter). As anchor tenant, LunaCorp has made a firm commitment to ISE that they will use ISE's services and pay a set price for them. ISE can then use this agreement in presentations to other potential clients, to demonstrate that the delivery service will actually take place, and that there is a demand for such services. This, theoretically, will convince those other prospects to sign on for the service, thereby guaranteeing the financial success of the company.

ISE currently has eight missions planned, including the initial orbiter and the LunaCorp mission. Some of the missions will land on the never-before-explored far side of the Moon.

These missions will put a number of payloads on the lunar surface, to be used in a variety of capacities. Scientists will finally be able to begin lunar-based astronomy. Free of Earth's limiting atmosphere, and with the solid ground of the Moon to build upon, telescopes of heretofore unimagined clarity and range will be deployed, undoubtedly changing our view of the universe. Robotic explorers will give geologists ample opportunity to test and refine theories of lunar composition and origin. Those same rovers, mounted with television cameras, will afford a unique educational opportunity, allowing students all over the world to explore the Moon first hand. Some students will even get a chance to operate a rover telerobotically, using the latest techniques of virtual reality. In addition, ISE intends to film and document the whole adventure, providing dramatic footage for television, movies, and videocassettes. ISE hopes all this will reignite general public interest in space exploration. According to Vice President Bell, "We're creating a new market, not competing with an existing market. The Moon is not [yet] a commercial market. It's bringing private investment into space."[4]

Notes:

1. International Space Enterprises (ISE) press kit, 1993.
2. *The Huntsville Times,* Huntsville, AL, September 29, 1993 (reprinted in ISE press kit).
3. ISE press release, September 29, 1993.
4. *The Huntsville Times,* September 29, 1993 (reprinted in ISE press kit).

Sources:

Space News, October 18-24, 1993, p. 29.

Four

LunaCorp

If the Moon is to be developed as a new arena for human activity, it will not be done by the government, according to the founders of LunaCorp. Rather, private enterprise will drive the expansion of the human economy. To this end, LunaCorp is embarking on an aggressive marketing campaign to generate enthusiasm for its Lunar Rover, to be landed on the Moon in 1997.

The President of LunaCorp is David Gump, who wrote a book in 1990 entitled *Space Enterprises: Beyond NASA*. In that book, Gump posits that NASA is incapable of opening space for the average person. If humans are to travel to space in any significant numbers, the effort must be spearheaded by the profit motive. "Space exploration has been reserved for government employees for too long," Gump said in the press release announcing the rover. "This will open exploration to everyone."[1]

Gump acted on his principles with the founding of LunaCorp in 1989. Until recently, the company concentrated on developing and marketing space oriented CD-ROMS, but with the announcement on February 14, 1994 that the company will build a lunar rover, it has moved into high gear in the quest to commercially develop space.

The lunar rover will be lofted to the Moon on International Space Enterprise's (ISE) first mission to land on the surface in 1997. The rover is slated to make a 1000-kilometer journey that will pass by and pay reverent homage to the Apollo 11 landing site (the scene of the first men on the Moon), followed by the sites of Surveyor 5 and Ranger 8 (both robot explorers), Apollo 17 (the scene of the last men on the Moon), finally searching for Lunakhod 2, a Russian Moon rover whose whereabouts are not precisely known.

The rover, whose development is perfectly on schedule, is being designed and built by Dr. William L. "Red" Whittaker and his team at Carnegie Mellon University's Robotics Institute. Dr. Whittaker was given funding from NASA to develop lunar rover technology, and was encouraged to seek cooperation with a commercial enterprise to use the technology on a commercially-financed lunar mission. Dr. Whittaker chose LunaCorp to carry out that mission.

Dr. Whittaker is on the cutting edge of robot development. His team constructed Dante, a semi-intelligent robot that was programmed to descend into Mount Erebus in Antarctica, to test operations in harsh conditions resembling the Moon or Mars. Due to technical difficulties, the robot completed only twenty-one feet of its descent. Dante II, however, is in development, and the information that Dr. Whittaker's team is gathering in these experiments will culminate one day in a teleoperated Martian rover. LunaCorp's lunar rover is an integral part in this process.

Gump has assembled several space luminaries on LunaCorp's board of directors. Chairman of the Board is Thomas F. Rogers, who is a co-founder of External Tank Corporation. Also on the board are Rick Tumlinson, President of the Space Frontier Foundation; Dr. William C. Stone, instrumental in the development of life support technologies; Scott Carpenter, a Mercury astronaut who also participated in an undersea habitat experiment; and Philip E. Culbertson, once a general manager at NASA, now Senior Vice President of External Tank Corporation.

The lunar rover mission is to be privately, commercially financed, and its activities and data are to be as widely disseminated as possible, to allow the general public to experience the thrill of exploration. According to Gump, "The costs will be paid by the people who are most interested in space. Visitors to theme parks, television viewers and contest entrants will be the primary funding sources, instead of taxpayers."[2] LunaCorp is in final negotiations with a theme park, which will allow visitors to drive the rover, and a television network, which will carry coverage of the rover's journey. The company has plans for extensive marketing to help pay for the project and to demonstrate that space activities have a market value. It will produce *The Moon Crew*, a children's education program based on the rover's activities, which will stimulate interest in learning about science and space. Information generated by the rover will be used in constructing "virtual reality" attractions at theme parks. It will sell the rights to a corporate sponsor to place its logo on the rover, and to use the data and images in its advertising. As the commercial viability of the project is demonstrated, more companies will want to participate, thus creating more ideas and incentives for marketing possibilities.

The four-year campaign which culminates in landing the rover on the Moon also offers a series of near-term events, in which sponsors will derive immediate returns on investment as these events draw attention and build public enthusiasm for the project:

- Third quarter 1994: Dante II descends into Mount Spurr in Alaska, with high media visibility.
- Fourth quarter 1994: The design of the lunar rover is unveiled.
- First quarter 1995 : Prototype rover begins making personal appearances for commercial sponsors.
- Third quarter 1995: Prototype rover conducts 100-kilometer trials on Earth.
- First quarter 1996 : Prototype rover begins telepresence demonstrations at exotic Earth locations.
- Third quarter 1996: Prototype conducts a 1000-kilometer trial on Earth.
- Fourth quarter 1997: Lunar rover lands on the Moon and begins MoonTrek 97.
- 1998 and beyond: Rover explores the Moon, transmits television images, carries out science research, and entertains visitors at theme park attraction.

LunaCorp's plans for commercial space projects do not stop there. The company intends to expand its activities until it can offer human space activity. When that day arrives, space will finally be opened up for average people to go work and play in space.

Notes:

1. LunaCorp press release, February 14, 1994.
2. LunaCorp press release, February 14, 1994.

Sources:

Space News, February 14-20, 1994, p. 17.

Five

The Artemis Project

The Space Program has lost its grandeur, according to Greg Bennett. "Mercury, Gemini, and Apollo carried human beings to the moon in less than ten years, but manned space programs lost sight of their exploratory nature and fell into lethargic bureaucracy when they adopted more pedestrian names like Skylab, the Space Shuttle, and the Space Station."[1] Bennett intends to correct that problem, and reinfuse space exploration and development with a sense of its mythic proportions. He has launched The Artemis Project, with the goal of a private, commercial, 100% self-sustaining Lunar colony. "Artemis is the twin sister of Apollo in Greek mythology," Bennett explains. "She is the moon, and Apollo is the sun. Artemis is also the goddess of the hunt, a constant reminder that our project is a voyage of exploration, a venture which will live off the land in its travels and return products of great value to our home on Earth."[2]

The Artemis Project, which is managed by the Lunar Resources Company, intends to be fully commercial, fully private, and economically self-sufficient from the first launch. Bennett and the other founders intend to film the whole project and release it as a major motion picture, thereby generating hundreds of millions of dollars. Bennett believes that the audience who once thrilled at the Apollo flights will pay to watch the drama of a lunar base under construction. The film, combined with other marketing ventures, including sales of scientific and technical data, videotapes, toys, games, moonrocks, gemstones, books, clothing and other promotional items, and television advertising revenues, will generate $1.34 billion. Such revenues will more than offset the $1.27 billion projected development costs of Artemis.

The culmination of the project will be an operating Lunar base which will commence permanent human presence on the Moon, exploit lunar resources for profit, demonstrate that manned spaceflight is within the reach of private enterprise, and "bootstrap" private industry into human spaceflight. The company has developed a timetable for the project, which Bennett, hoping to avoid one of NASA's common pitfalls, is quick to remind us "is a working schedule, not a worshipped schedule":

- 1994-95: Feasibility study and exploratory design.
- 1996-97: Preliminary design.
- 1997-2001: Final design, development, and testing.
- 2002: First flight.

The company has already plotted out the Artemis Reference Mission, which is being used in the preliminary feasibility study to determine costs, revenues, and technical and political issues. The reference mission calls for the launch of mission components on two Space Shuttle flights, which will then be assembled in orbit. The company has opted for the Shuttle because, although the most expensive of current launchers, it is currently the most capable launcher certified to carry a crew. Bennett and his colleagues are aware of the new launch vehicles under development, some of which have been detailed in this book, but none have progressed to the stage where they might be considered for such a mission. Should any of them achieve operational status within the time frame of the mission, the company will consider them as launch alternatives.

The mission is comprised of three elements. The Lunar Transfer Vehicle will carry the crew to and from the Moon. The Descent Stage/Lunar Base Core Module will take the crew to the Lunar surface and provide a pressurized habitat for Lunar operations. The Ascent Stage will return the crew after the initial mission to the orbiting Lunar Transfer Stage.

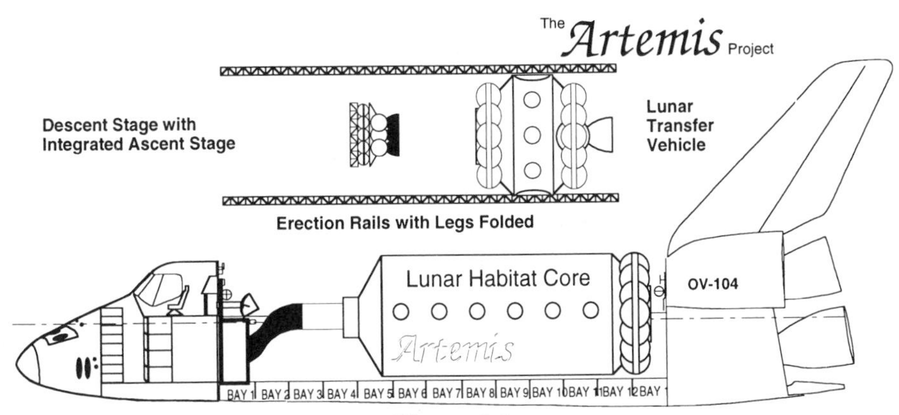

Figure 5-1

Stage One of the Artemis Mission involves transporting a Lunar Habitat Core to the Moon using the Space Shuttle or similar vehicle. (Illustration courtesy of The Artemis Project.)

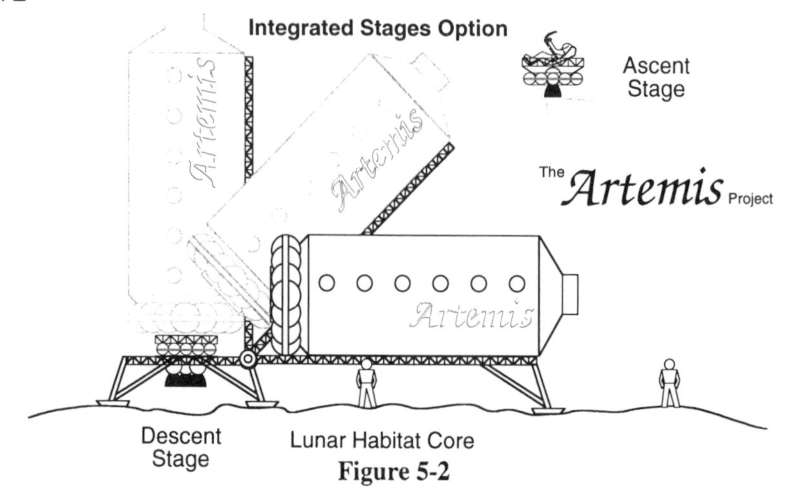

Figure 5-2

In the Descent Stage, the Lunar Habitat Core is lowered to the surface of the Moon and configured for use. (Illustration courtesy of The Artemis Project.)

The Artemis Project

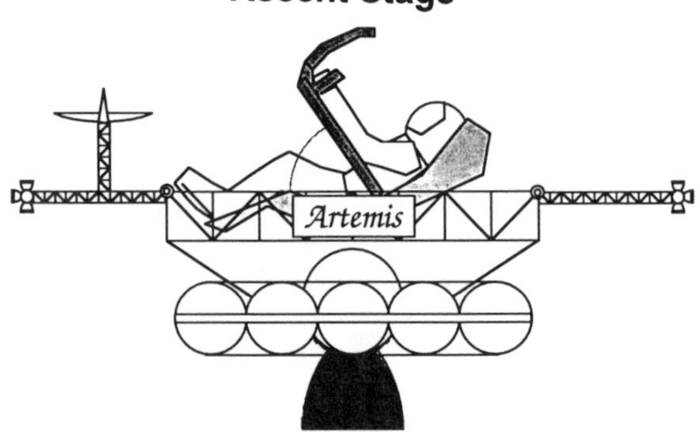

Figure 5-3

A small lunar shuttle is used to transport crew members from the lunar surface to the hovering Lunar Transfer Vehicle. (Illustration courtesy of The Artemis Project.)

After in-orbit assembly, the complete craft will be sent to the Moon. There, the other two components will separate from the Lunar Transfer Vehicle and descend to the Lunar surface. It is hoped that the Lunar Transfer Vehicle can be fully automated, which would eliminate the need for a pilot to stay on board. This would reduce the initial weight of the mission, and thus the cost.

On the Moon, the crew will ensure that the Lunar Base Core Module is level and will reconfigure it for Lunar operations. All the while, they will be mounting and operating cameras to record the whole venture. The film will later be edited for theatrical release. They will also engage in surveying and assaying the vicinity, and preparing for the next mission. When the goals of the first mission are achieved, the crew will ascend in the Ascent Stage to the waiting Lunar Transfer Vehicle, and the entire craft will return to Low Earth Orbit (LEO). There, the craft will rendezvous with the proposed International Space Station, and return to Earth on the Space Shuttle. The company has alternate plans for a mission that does not employ a space station, but such a design adds cost and complexity to the whole mission. If the Space Station exists at the time of the first mission, the company plans to use its capabilities to make The Artemis Project as efficient and inexpensive as possible.

Regardless of NASA's future developments, The Lunar Resources Company intends to run The Artemis Project as cost-effectively as possible, using the private enterprise paradigm. The company is currently working toward incorporation, and will sell stock as soon as is practical. Bennett has concluded that:

> [a]nalysis of government-sponsored space shows that no more than 10%, usually even less, of the money is actually spent on developing and operating the

> spacecraft. The rest goes to the enormous support effort and inefficient organizations necessary to answer the changing whims of the U.S. Congress, support a large institutional bureaucracy [NASA] with extensive fixed assets all over the world, and to adapt to the government's management-by-meetings philosophy... Private enterprise does not, and could not, work that way... [T]he costs of any program can be reduced by a factor of ten or more.[3]

The challenges are great for The Artemis Project, and many uncertainties yet exist, such as the availability of the Space Shuttle or the Space Station for participation in the project, or if NASA will even talk to the company after such harsh analysis. If Bennett can pull it off, however, he will become heir to the title currently held by the Robert Heinlein character D.D. Harriman, who is "the Man Who Sold The Moon." In fact, Bennett says many people ask him if he thinks he is Harriman. He replies, "Nope; Harriman had it easy. More like P.T. Barnum."[4]

Notes:

1. The Artemis Project, Frequently Asked Questions, Revision 8, May 19, 1994.
2. *Ibid.*
3. *Ibid.*
4. *Ibid.*

Six
OUSPADEV

"Imagine what it would be like to take a vacation in space ... playing in near-zero gravity ...donning a space suit and venturing out into the void of star-studded blackness ...the fun of discovering how to do everyday activities like eating, or drinking, all while orbiting high above our planet ...watching the sun rise and set every 90 minutes ...sleeping (and other after-dark activities) in weightlessness ...or just sitting and watching the cosmos drift by the viewport in utter silence."[1]

That is the vision of OUSPADEV, The Outer Space Development Company. The company intends to provide such vacations by the year 2002 on its private space station *Prelude*. The station, which is the first privately planned enterprise of its kind in the world, is to be financed by a series of stock offerings. A private space station funded by private means, hopes John House, President of OUSPADEV, will break

NASA's stranglehold on access to space and make space flight and vacations available to everyone.

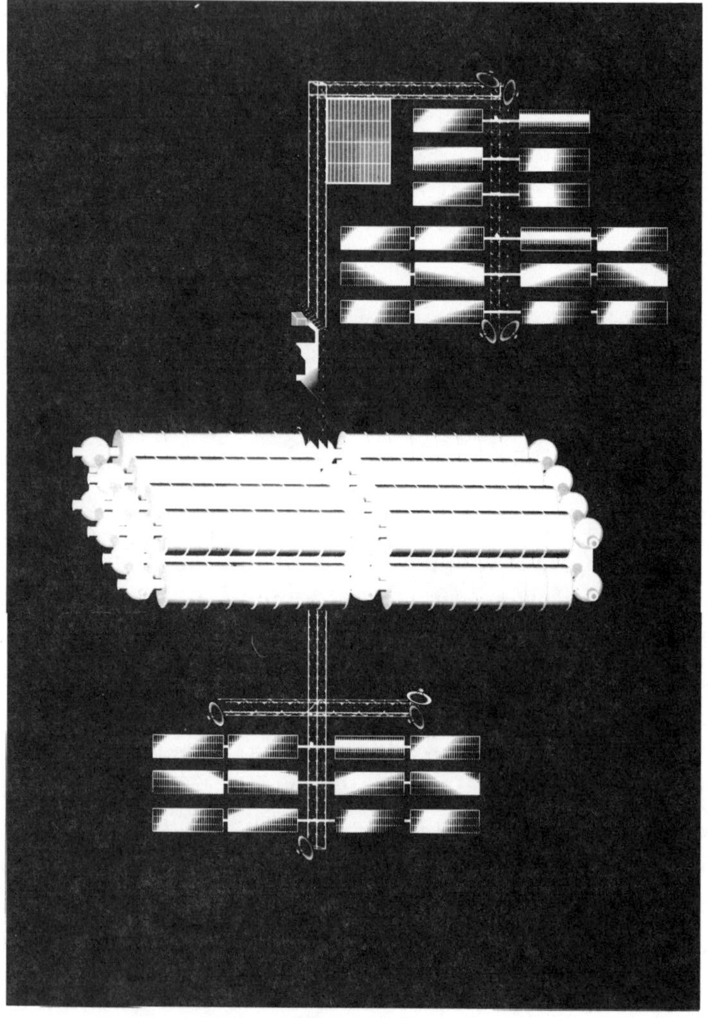

Figure 6-1

Computer-generated image of the proposed OUSPADEV space station. (Illustration courtesy of The Outer Space Development Company.)

Vacations in space will be a central theme of *Prelude*, but House knows that he will need more than tourists to foot the bill. OUSPADEV's development plan calls for leasing twelve of the 16 modules of the station to corporations, who can use them for any space applications that are not illegal and would not damage the station. Each module, containing 11,000 cubic feet of usable volume, would be leased on a five-year basis for $1 billion. OUSPADEV would provide basic services such as power, life support, housing, medical care, and recreation for a company's staff. House has identified at least 200 companies with the financial strength to afford a lease that are already involved in, or could benefit from, space research and development.

The other four modules of the station will be reserved for station operations and the vacation facilities. Activities will include space sports, flying under one's own power on wings strapped to one's arms, Earth and star observations, and, for the romantically inclined, zero-G love grottoes.

In a recent press release, House said:

> The space program costs so much today for two simple reasons, greed and more greed (disguised as management). The companies that work with NASA and with the Department of Defense only have one real client: the U.S. government. The only way they can increase their sales/profit is by increasing the amount of money the government spends. And since it is against unspoken government policy to ever reduce the amount of money spent, a perfect marriage was born and has lasted for decades.
>
> OUSPADEV, however, is market driven. We are interested in getting the best product at the lowest price. Because of that, we use existing technology.

> Why reinvent the wheel? OUSPADEV also eliminates layers upon layers of government and big business bureaucracy. OUSPADEV is a simple business building a straightforward product. There is a demand in the market, we supply the product.[2]

After a decade of market research, House founded OUSPADEV in 1991 and proceeded with the plans for *Prelude.* House is joined in the effort by William Morrow Tracer, a computer systems designer who will develop *Prelude's* computer hardware and software. They are aided by marketing director Sherman Robbins, CEO of Robbins and Ries, Inc., a communications holding company, and Anthony E. Häag, who will build OUSPADEV's Orbit Club into a worldwide space enthusiast network.

The first stock offering, in June 1995, will raise $5 million, enabling the company to proceed with designing the station. House estimates the station will cost $6-10 billion, and will take approximately ten years to reach full operational capacity. In the next twelve years, the company is projected to earn an after-tax profit of $7.5 billion on an investment of $1.165 billion and sales of $16 billion. The station is expected to be operational for 25 years. If all goes as planned, the station will be built according to the following timetable:

- 1994-95: Obtain design financing through stock sales. Begin assembling design team.
- 1995: Complete station design and build model/prototype.
- 1996: Sign first lease contract.
- 1997: Sign two additional lease contracts.
- 1998-2000: Sign remaining nine contracts.

- 2000: Launch first space station components.
- 2001: Christen station — rudimentary operations.
- 2002: Station fully operational — invite first vacationers aboard.

Until the day that *Prelude* is open for business, there are several ways OUSPADEV is promoting private space development. The company runs a Vacation In Space program, wherein would-be space tourists can invest a regular monthly amount in an interest-bearing account toward their first orbital vacation. Another organization, The Orbit Club, is designed to bring together space enthusiasts in their pursuit of privately-developed space access for everyone. House is currently seeking more members, especially to head local chapters. The Club Catalog contains a compilation of merchandise that is of interest to Orbit Club members and space enthusiasts in general. *Outward*, a quarterly newsletter, keeps members up-to-date on the club's activities, and on the state of commercial space. The company maintains a computer bulletin board, with regular updates on *Prelude*, and on commercial space in general. Finally, OUSPADEV hosts The Space Awareness Expo annually in Fort Lauderdale, Florida. The third annual conference, held on the weekend of July 29-31, 1994, featured three seminars on the state of space development. One was concerned with astronautics, one with launch vehicles and concepts, and one covered space business developments.

If, by 2002, you can take a vacation in orbit aboard *Prelude*, you'll have the foresight and vision of John House and his associates at The Outer Space Development Company to thank for it. It will no doubt be the first step to a booming space tourism trade which could eventually produce Disneyland-like theme parks on the Moon, or Earth-orbiting space casinos. Space will then truly be open to anyone who can scrape together the cash equivalent of cruise ship vacation today.

Notes:

1. OUSPADEV press kit.
2. *Ibid.*

Seven
Kistler Aerospace Corporation

Bob Citron wants to go to space. He wants to bring everybody else along as well. That is why, in 1980, he founded The Space Travel Company, Inc. Because the infrastructure for space tourism was not yet in place, that company did not make it, but Citron has been developing his space tourism ideas ever since. "I have always believed that space tourism will become *the* (not *a*) dominant factor in the development of space during the coming century,"[1] he wrote in a recent fax communication.

Citron is currently working on creating the necessary infrastructure for space tourism. He is president of Kistler Aerospace Corporation (KAC), which is building the world's first privately-funded single-stage-to-orbit (SSTO) vehicle. He has experience in bringing seemingly far-out space concepts to fruition. In 1983 he founded Spacehab, Inc., to construct a module that rests in the Space Shuttle's Middeck and carries commercial experiments. Spacehab was developed with private funds, and currently holds a contract with NASA for several flights aboard the Shuttle, two of which have already flown. Spacehab is a success, but not exactly in the way Citron envisioned it.

Citron originally planned Spacehab to be a passenger-carrying module that would ferry tourists into space for some zero-G fun. When it became clear, however, that NASA was never going to let non-astronauts fly on the Shuttle, Spacehab changed its focus to the experiment module. Citron oversaw Spacehab's development to the point that it transitioned from the conceptual stage into hardware construction, then stepped down from his position as Chairman of the Board in 1988. He still burned with the idea of jump-starting the space tourism industry, and turned his attention to bringing it to pass.

With his original Spacehab partners, he founded Kistler Aerospace Corporation in the fall of 1993. Co-founders include Tom Taylor, who also runs Global Outpost, a company dedicated to turning spent Shuttle external tanks into space stations, and Walt Kistler, who is head engineer of the company's launcher efforts.

KAC intends to offer regular passenger service aboard an SSTO craft to Low Earth Orbit (LEO) by 1998, and has a step-by-step plan of how to achieve that goal. The first step is to build and fly a sub-scale demonstrator, dubbed the K-0. This craft will stand 12 feet tall and will be test-flown in the Mojave desert at the end of 1994. Using data from the demonstrator, the company will design, build, and test the K-1, an SSTO craft that stands 35 feet tall and will be capable of lifting 2,000 lbs. to LEO. Development of the K-1 will take place during 1995 and 1996, with first flight by the end of 1997. The K-1 will begin normal operations in 1998, and will be joined by a fleet of similar vehicles.

Concurrently, and using data from the K-1 program, Kistler will design and build the K-2 launcher. This craft, which will be operational by 2002, will have a height of 85 feet, and will be capable of placing 20,000 lbs. into LEO. The company will build a fleet of K-2s as well. To support these activities, KAC is planning a launching and operations facility as part of the recently-announced White Sands, New Mexico commercial spaceport. Construction will begin in 1996, and the spaceport will be able to accommodate the K-1 by 1997, and the K-2 by 2001.

Citron estimates it will cost from $1 billion to $2 billion to fully complete the system in a decade, but he foresees the ships returning many times that in a burgeoning space economy. KAC has had some success in raising the necessary capital, but Citron declines to elaborate. The K-2 spacecraft will actually have several different configurations, suited to different tasks, including crew and cargo transport to and from orbital facilities, space tourist passenger transport, placing larger satellites and other payloads into Earth orbit, and as orbital tanker fleets that will become part of the space infrastructure for geosynchronous and lunar operations. In addition, Citron forecasts many new, unpredictable space-based businesses will crop up, stimulated by the reduced cost of access to orbit.

The keys to Kistler's launcher concept are reusability and streamlined, air-carrier-like operations. "Commercial transportation systems are economically viable and most profitable when two things happen," Citron writes in the company's business proposal. "1) after completing a job, the delivery vehicle is put back into service quickly, and 2) the ratio of labor costs to hours in service is kept low." The Space Shuttle and the U.S.'s fleet of expendable launch vehicles do not meet these criteria, according to Citron.

> The bottom line is that currently available delivery systems are far too expensive, limiting access to space to a few wealthy companies and governments. These prohibitive costs are, in effect, costing our country future economic growth and prosperity.... Kistler rockets will open up the space frontier for humanity by reducing the cost of access to space from thousands of dollars per pound to just a few hundred dollars per pound.[2]

Moreover, the Delta Clipper vehicle being built by McDonnell Douglas does not compare favorably to Kistler launchers. Delta Clipper development will cost $5.5 billion, all taxpayer money, and will not fly before 2004. The K-1 is completely privately financed, will cost $300 million for three ships, and will be operational by 1998. The K-2 will cost $2 billion for 4 vehicles (again, all private money), and will fly by 2002. Moreover, since the Delta Clipper is completely paid for by the government, there will probably be as much likelihood of commercial passengers flying on it as on the Shuttle.

The Kistler vehicles, on the other hand, will be commercial ships from the first flight. Citron hopes to use them to develop space tourism as quickly as possible. He envisions nine phases of tourism, beginning with quick jaunts to space, and finally resulting in a Lunar Surface Resort. Initially, tickets will sell for $50,000, but that price is not far out of line with high-end, earthbound adventure travel and cruise ships. As the K-1 and K-2 fleets are augmented and flight frequencies increase, however, the price will drop dramatically, to somewhere around $25,000 for a stay at the lunar facility. Citron's nine phases build upon one another:

1. 2005: Space-available seats on routine space station crew/cargo transfer flights (to *Prelude*, perhaps?).
2. 2006: One-day LEO trips that co-orbit, but do not dock, with a space station.
3. 2010: One-day LEO trips that dock with a space station.
4. 2012: Three- to seven-day trips that dock with a LEO facility. Tourists engage in astronomical and Earth-observation activities.
5. 2015: Fifteen- to twenty-day trips to a LEO space facility. Tourists engage in amateur research, or participate in the station's programs.
6. 2018: One- to four-weeks aboard a LEO Space Resort Facility (SRF). Tourists engage in microgravity demonstrations, gymnastics competitions, extravehicular activities (EVAs), space and Earth observation astrodomes, space "entertainment."
7. 2020: Ten-day tour of low lunar orbit. In addition to the SRF, tourists orbit the Moon several times.
8. 2025: Two-week lunar base visit. Tourists engage in short lunar surface excursions.

9. 2030: Two- to three-week visit to a Lunar Surface Resort (LSR). Tourists engage in extensive lunar exploration and seminars on lunar science.

When the space age dawned, many people who had yearned for space travel believed the day of their personal flight into space was right around the corner. Almost forty years later, we are still waiting. If Bob Citron and Kistler Aerospace Corporation can unleash the power of market forces on the problem of the average person's access to space, we will get our chance, in less than a decade, to frolic in the playground of open space. The long wait is almost over.

Notes:

1. Personal communications, February 12, 1994.
2. Kistler Aerospace Corporation business proposal, May 2, 1994.

Eight
Hudson Engineering

Gary Hudson has been trying for several years to build a commercial space vehicle that will climb out of Earth's gravity well on a single stage. Yet due to the resistance he has encountered from the status quo, he has been struggling at the bottom of a deep gravity well of ideas and perceptions, with the odds stacked against him. His design, however, is valid, and the philosophy driving it is so compelling that he will not give up.

Hudson proposes to build an air-launched, horizontal landing SSTO (single stage to orbit) vehicle, using only private funds, for commercial applications. He has been involved in the SSTO scene for a couple of decades, and wrote an exhaustive history of the design concept, so he knows whereof he speaks. He wants to recapture the spirit that infused Apollo, but has since been frittered away, according to him. "There is no permanent human presence in space, only the promise of a space station that even supporters doubt will be in place by the end of the century.... This is not the future we imagined or hoped for when we watched Neil Armstrong make that first small step in 1969."[1]

The frontier of space will never be opened if left to the government, Hudson believes. Just as the American frontier was opened by private citizens in covered wagons, so will space be opened by pioneers in private spaceships. The proper role of government is to provide legal, military, and police protection, but not excessively. Yet the government has taken over the space enterprise, closing out private citizens. "It has such a stranglehold on space development that most people genuinely believe space activities are too complicated and too costly to be trusted to any organization outside of NASA and the bloated aerospace corporations."[2]

Hudson firmly believes that the space frontier must be opened, if we are to progress with any degree of certainty into the future. "We want to go to space because, quite simply, it is the next habitable frontier."[3] That is all the more reason to go to space for Hudson. No amount of secondary reasoning will make the argument more powerful. All other reasons — scientific research, promoting world peace, cooperation with the Russians, uplifting the human spirit — will develop as a natural result of moving into the next frontier.

The current obstacle to settling that frontier, as Hudson sees it, is the cost of getting there. Current Shuttle and expendable launch vehicle costs run anywhere from $1500 to $3000 per pound, he estimates. At those prices, we'll be stuck on the ground for a long time. He believes if we can lower those costs, there would be an economic boom unrivaled in history.

Hudson has a solution. He has designed and is seeking funding to build a variety of SSTO craft. An SSTO vehicle, run with airline-like efficiency, can bring launch costs as low as $100 per pound. At those prices, current space businesses like communications satellites, and as-yet undeveloped concepts such as space tourism, will become eminently more affordable.

The Rocketplane is Hudson's first goal. Once he receives the necessary funding, it could be operational within three years at a cost of around $20 million. His company, Hudson Engineering, has already committed several hundred thousand dollars toward vehicle design and subscale model construction and testing. The Rocketplane will carry two crew members and two passengers, plus a little cargo, to LEO. He plans a bigger version, called the Spaceplane, which will carry five to eight tons to orbit, or as many as forty passengers. The propulsion and avionics of the two vehicles are essentially identical; only the airframe will be different. The Spaceplane can be ready for flight about three years after the Rocketplane, for a development cost of less than $100 million.

Hudson's goals for the craft are complete reusability and airline-like operations. Any commercial airport anywhere in the world will be able to support Rocketplane and Spaceplane flights, according to him. All that is required is a runway of at least 8,000 feet, and access to cryogenic tankers for the delivery of the fuel, consisting of liquid oxygen and liquid hydrogen. He estimates the direct cost of operating the vehicles at about $50,000 per flight for the Rocketplane, and around $200,000 for the Spaceplane, resulting in a per-seat cost of about $5,000, or $20 per pound to deliver cargo to LEO.

Not content with just one or two options to get to space, Hudson has teamed up with Bevin McKinney of American Rocket Company to design the Roton, a craft similar to a helicopter, but with rockets on the ends of the blades. This craft has three distinct advantages over existing launchers. First, it is totally reusable, and has safe-abort capability which will minimize the chances of accidents. Second, it is small and uses off-the-shelf technology and inexpensive fuel. Finally, its development can be carried out by a small team of less than a dozen engineers and technicians, which can make extensive use of subcontractors.

The Roton design team believes that the craft will economically launch satellites to orbit, as well as open up the space tourist trade. The vehicles will be developed, manufactured, and sold to operators, just like airplanes are, thus taking advantage of a familiar and well-known business paradigm.

The Roton would take off like a helicopter, with the rockets at the tips of the blades firing as well, at an angle to the ground. The lifting capacity of the whirling blades is added to the rockets' thrust. Once the blades can provide no more lift against the atmosphere, the rockets are turned straight down, and the craft achieves orbit under rocket power. On reentry, the blades act as an aerobrake to slow the craft down, as well as giving it the ability to guide itself through the atmosphere. Roton vehicles would be able to take off from and land on any level ground, minimizing support infrastructure.

Hudson believes these and other reusable spacecraft concepts, built using the principles of private enterprise rather than government contracting, are the key to opening the space frontier for average people. He further believes that the time is ripe for such craft, and that the first entrepreneurs to develop such vehicles will profit immensely. "It is a challenge, but one we can neither resist nor refuse," he asserts. "We were there when Apollo landed on the Moon. We remember the promise of the future we envisioned. We have reached a pivotal point in history. We are the last generation who personally witnessed man on the Moon. For those who follow us, NASA has already written the history books on why we didn't go back. If man is to settle the frontier of space, brave, far-thinking individuals must take on the task themselves."[4]

Mars Aerial Platform (MAPS) hovering above the planet will be used to take detailed measurements of the Martian environment. From the Mars Direct Project. (Photo courtesy of Robert Murray, Martin Marietta Astronautics.)

The Mars Hopper will be used for ground transport by Mars Direct. (Photo courtesy of Robert Murray, Martin Marietta Astronautics.)

The "Rocketplane" will be able to explore vast areas of Mars.
From the Mars Direct project.
(Photo courtesy of Robert Murray, Martin Marietta Astronautics.)

The proposed base from the Mars Direct program can
also be used to colonize the Moon.
(Photo courtesy of Robert Murray, Martin Marietta Astronautics.)

Notes:

1. Hudson, Gary, C., "Common Sense Spaceships," June 8, 1992, p. 1.
2. *Ibid.*, p. 2.
3. *Ibid.*, p. 3.
4. *Ibid.*, p. 9.

Sources:

Personal communications, September 21, 1992.
Personal communications, May 10, 1994.

Nine
Mars Direct

In 1805, Lewis and Clark set out to explore the territory President Jefferson had just acquired under the Louisiana Purchase. After months of living off the land, staying with various Native American tribes, and employing them as guides, they reached the Pacific Ocean.

In the later Nineteenth Century, Roald Amundsen and his team of five succeeded in forcing the Northwest Passage, north of Canada, where many larger, well-equipped British Navy expeditions consisting of several large ships failed, with many men even perishing. His secret: living off the land, like an Eskimo (albeit with the aid of modern inventions like rifles and skis).

In 1953 Dr. Wernher von Braun proposed "the Mars Project." His scheme involved 10 spaceships laden with everything necessary for the voyage to Mars, not unlike the British Navy expeditions in the Northwest Passage. His plan required a complex space station in orbit around Earth for construction of the spacecraft, and a crew to remain in orbit around Mars while another contingent explored the planet.

In 1989, President George Bush heralded the Space Exploration Initiative (SEI), a grand, von Braun-esque vision of returning to the Moon and pressing on to Mars by the year 2019, the fiftieth anniversary of the first human footfall on the Moon. The cost: $400 billion (at least). The United States Congress choked on the figure, and canceled all funds for the plan.

Robert Zubrin, an engineer at Martin Marietta Astronautics near Denver, scoffs at such a drawn-out plan with such a ludicrous price tag. "[I]f Kennedy had said, 'Let's go to the moon by the year 2000,' his support would have been lukewarm at best," he said in a *Smithsonian Air & Space* magazine interview, "and things would have fallen apart long before. Supporting a 'gradual approach' is just another way of saying you don't want to go."[1] In his opinion SEI takes too long, costs too much, is overly complex, and doesn't accomplish anything. "It is not enough to go to Mars," he believes, "it is necessary to be able to do something useful when you get there. Zero capability missions have no value."[2]

Zubrin has a much cheaper, much more effective plan in mind. Dubbed "Mars Direct," it could put humans on Mars as early as 2001, facilitate extensive exploration, and create an infrastructure for future human presence on Mars, eventually leading to permanent settlement. Total cost: $40 billion, plus $1 billion for each launch in the program. Mars Direct would accomplish the aims of SEI eighteen years sooner for a tenth of the cost. Moreover, upon completion of the mission, the seeds of the first off-world colony would be planted.

The secret of Mars Direct, as with Lewis and Clark and Amundsen, is to "live off the land." As many of the local Martian resources as possible would be used. Moreover, most of the technology for such a mission is already available; no need for expensive research and development programs to construct the necessary equipment. The "direct" in the plan's name means sending the expedition directly from the Earth's surface to the Martian surface. No need for complex space stations or interplanetary armadas.

Mars Direct calls for an average of one launch a year, starting in 2001. That is one-tenth of the U.S.'s launch capacity, which Zubrin maintains would be eminently sustainable. The first launch, named Ares 1 (Ares is the Greek name for the god of war, whom the Romans called Mars), launches a payload of 40 metric tons to Mars. This payload consists of an unfueled methane/oxygen driven two-stage Earth Return Vehicle (ERV), 6 metric tons of liquid hydrogen, a 100-kilowatt nuclear reactor mounted in the back of a methane/oxygen driven lightweight truck, an automated chemical processing unit, and a few small scientific rovers. This payload aerobrakes into orbit around Mars, then descends to the surface on a parachute (aerobraking is the process of using a planet's atmosphere to slow a craft down in preparation for landing).

Once on the surface, the truck is driven telerobotically (by remote control) a few hundred meters from the lander. When a safe distance is reached, the reactor is deployed and started up. This provides power for the chemical processing unit, which proceeds to manufacture propellant for the return journey. By a process that has been practiced on Earth since the 1890s, the hydrogen is reacted with carbon dioxide, which comprises 95% of the Martian atmosphere, to produce methane and water. The methane is liquefied and stored, and the water is electrolyzed into hydrogen and oxygen. At the same time, a different process breaks down the Martian carbon dioxide into carbon and oxygen. The original 6 metric tons of Earth hydrogen, combined with local Martian resources, produces 24 metric tons of methane and 48 metric tons of oxygen, while the other process generates 36 metric tons of oxygen, for a total fuel leverage ratio of 18 to 1. That is living off the land! A working scale model of the processing plant has been built by Zubrin and a team of Martin Marietta engineers.

While the plant is doing its work, the rovers scout the immediate area and relay the information back to Earth. Project managers on Earth monitor the propellant production and use the rover data to select the optimal landing spot for the crewed expedition. If all goes according to plan, two more Ares vehicles are launched toward Mars in 2003. The payload of one of the launchers is identical to the original payload. The second launcher, however, carries a habitation module (the "Hab") and a crew of four, along with food and supplies for a three-year stay, a pressurized ground rover, and an aerobraking/landing structure. Once the crewed craft is on its way to Mars, a tether is extended between the Hab and the spent second stage of the launcher, and the whole assembly is set to rotating. This generates a centrifugal force which produces an effect similar to the gravity found on the Martian surface. The crew spends half an Earth year in transit to Mars, continuing their training for the mission.

Upon arrival at Mars, the Hab aerobrakes and descends to the surface. Most likely everything will go as planned (in the Apollo program, only one out of ten crewed flights experienced difficulties), but if the Hab should miss the preselected landing site, there are three backup contingencies. First, if the Hab is under 1000 kilometers off the mark, the pressurized rover will have sufficient fuel to drive to the fuel processing plant. Second, if the Hab lands over 1000 kilometers away, the second fuel production plant, which was concurrently launched, can be landed near the crew. Finally, if that plan fails, the crew has provisions for three years, long enough to mount a rescue mission from Earth. If there are no problems, however, the crew begins its year-and-a-half exploration of Mars, and the second processing plant is landed 500 kilometers away, in preparation for the next crew, to be launched in 2005.

The crew will be very busy during its 500-Earth-day stay. Zubrin believes there are many activities that can only be carried out by humans, as opposed to those scientists who hold that robots can adequately explore Mars. While robots are excellent for initial surveying and identifying the gross characteristics of Mars, humans will be required to do the detailed follow-up work, and to make decisions based upon the robotic discoveries. The most important research the crew will perform will be the search for Martian life. Current evidence is inconclusive on the question, and human presence is necessary to properly assess the data, and formulate revised search plans based upon newly-acquired evidence. If life is found, Zubrin believes, it would indicate that life on Earth is not a fluke, and that the likelihood of life throughout the universe would be overwhelming. Moreover, the discovery would lead us to a greater, more detailed understanding of what exactly life is, possibly leading to advances in medicine and disease prevention.

In addition, the crew will engage in a detailed survey of the vicinity, geologically characterizing the Martian landscape. The crew will use the pressurized rover to explore the surroundings of the Hab. Ten percent of the fuel produced by the processing plant is allotted to the rover, allowing it to cover over 22,000 kilometers of Martian territory. They will search for, among other things, a local supply of hydrogen. If it can be found, for example, in water ice at the poles, the need to transport it from Earth disappears, and that six metric tons of payload capacity could be replaced with additional science equipment on future Ares launches. The crew would also engage in experiments in hydroponic agriculture under a domed, pressurized greenhouse, to see if a colony could ever become self-sufficient. Such experiments are critical if Mars is ever to be permanently settled, believes Zubrin. "The possibility of creating a new branch of human civilization on Mars depends primarily upon the ability of the Mars base (or bases) to develop local resources to support a significant population."[3] In other words, the settlers must lift themselves up by their own bootstraps.

Every two years, two more Ares vehicles could be launched, like the two planned for 2003, for as long as the planners of the mission cared to continue the project. Each processing plant would be landed 500 kilometers from the last one, in a string around the planet. Eventually, any one of those sites could become the nexus for the first permanent human colony on Mars. Modules could be added onto the original Hab, until a veritable Martian city was created. Zubrin hopes for a town of 100 settlers by the year 2030.

The planet's exploration could eventually be carried out by a craft Zubrin has designed called the Nuclear rocket using Indigenous Martian Fuel (NIMF). Each time it lands, the NIMF pumps Martian carbon dioxide (CO_2) into its propellant tank to a pressure of 100 pounds per square inch, whereupon the gas automatically liquefies. The vehicle then heats the fluid and expels it out its rocket nozzle, producing thrust. Such a vehicle could fly from place to place in a single "hop," refueling itself after each landing. A scaled-down version of Mars Direct could be used to set up a permanent base on the Moon as well.

Zubrin's vision does not stop with mere exploration of Mars. He dreams of one day modifying Mars so that it is inhabitable by humans without space suits or oxygen masks. Such "terraforming" would take many years, if not centuries, but it is not outside the capabilities of our technology. The same process that has made Venus the local equivalent of hell and which threatens our planet can be used to benefit humans on Mars. The greenhouse effect is the process by which carbon compounds, mostly chlorofluorocarbons (CFCs), trap the heat of the sun within the atmosphere as it is reflected off the surface of the planet, thereby raising the mean temperature of the surface. On Earth, this process threatens to melt our polar icecaps and flood our coastal cities, but on Mars, it can be used to raise the temperature so that water can once again freely flow. This process eventually becomes self-perpetuating in what is known as a "runaway greenhouse effect," raising mean planetary temperatures by tens or even hundreds of degrees Centigrade (C).

Such an effect on Mars would be of great benefit to humans. (The natives, however, might complain; we should conduct a full survey and generate a full catalog of Martian characteristics before we engage in such a scheme. If we do find life, the ethical questions about creating a greenhouse effect become very complicated.) A factory could be constructed on Mars that would generate a certain amount of CFCs, which would begin to trap the Sun's heat and raise the mean temperature of the planet. As the temperature increases, the polar caps, which are mostly frozen CO_2, would begin to melt. The CO_2 would be released into the atmosphere as a gas, where it would act as an insulator that would contribute even more to the global warming. A rise in the mean temperature of Mars of 4ºC could cause a runaway greenhouse effect raising the Martian temperature 55ºC. This would make it warm enough, during the summer months, to melt the water ice which is trapped in the polar caps and in the ground. Water would once again flow on Mars!

Eventually this process would lead to a thicker atmosphere, in which humans could walk around without a spacesuit, although they would still need oxygen masks. It would be a few degrees too cool, however, for terrestrial plants to thrive. To truly terraform Mars, Zubrin has concocted a more drastic plan. He wants to find an asteroid out among the gas giant planets and redirect it so that it slams into Mars. Current astronomical theory maintains that there are numerous asteroids orbiting in the vicinity of Saturn, which are probably composed of frozen gases. A few well-placed nuclear thermal rocket engines could redirect such an asteroid so that it would make a close swingby of Saturn. This gravitational assist could be used to send the asteroid on its way to Mars. Thirty years after the initial redirection, the asteroid would crash into Mars, exploding with the force of 70,000 one-megaton hydrogen bombs (but without the radioactivity). Forty such bombardments would double the nitrogen content of the Martian atmosphere and throw huge amounts of material into the sky, thus boosting the greenhouse effect. At the conclusion of the onslaught, however, Mars' temperature and atmospheric pressure would approach that of Earth's, and humans could begin in earnest to make Mars their home.

Zubrin's grand terraforming schemes would require immense amounts of energy, more than the total available on Earth. Moreover, by the time such plans begin to become practical, the inhabitants of Earth will need much more energy than is currently available. "The history of our technological advance can be written as a history of increasing energy utilization," Zubrin asserts. "A century or so from now, nuclear fusion will be humanity's primary source of large-scale power production."[4]

Zubrin estimates that by the year 2200, humanity will need 2,500 terawatts (TW) of power per year (one TW equals one million megawatts), as opposed to our energy usage of 12 TW in 1992. He estimates the total power available from terrestrial fossil fuels at 3,000 TW, and the power available from nuclear fission with breeder reactors at 22,000 TW. If we are to continue to increase at our exponential rate of growth, especially if we are to colonize the planets, he maintains, we need a vastly greater amount of power.

Fusion, on the other hand, fueled by deuterium (an isotope of hydrogen, plentiful throughout the solar system) and helium three (^3He, an isotope of helium) can provide for our energy needs indefinitely. Zubrin surmises that there is enough ^3He in the atmospheres of the gas giant planets to provide 14 billion TW of power! The intense gravity of Jupiter, however, precludes the possibility of extracting the ^3He from its atmosphere, thus eliminating 5.5 billion TW of potential. That still leaves 8.5 billion TW of potential amongst the other gas giants, however, providing enough energy for anything humans could conceive in the next millennium or two.

Although fusion has not yet been demonstrated to release as much energy as it consumes, Zubrin believes that it is only a matter of a few decades at most before we will be able to use it for energy production (the Sun, after all, has been using fusion to throw off energy for five billion years). When that day arrives, the whole solar system will be open for human activity. Using thermonuclear fusion engines, voyages to Mars shrink from months to days, and outer solar system excursions can take place in months, instead of years. Operating at optimal efficiency, a fusion rocket can generate velocities up to five percent of the speed of light, finally bringing interstellar flight within humanity's grasp.

To mine the ^{3}He, Zubrin has designed the Nuclear Indigenous Fueled Transatmospheric (NIFT) vehicle, which would use a gas giant's atmosphere as fuel for a ramjet as it collected ^{3}He. The NIFT could also rendezvous with a floating ^{3}He production facility and collect its load. It would then take the load to an orbiting fusion-powered tanker which would transport the cargo to the inner solar system. The ^{3}He would be used as the fuel for the nuclear rockets that would redirect the asteroids out near Saturn toward Mars. If such technology comes to pass, the whole solar system would become the playground of humanity.

Today, looking forward into the future and out into the solar system, getting off the Earth seems a daunting task. If we "live off the land," however, like the early explorers, and pull ourselves off the Earth by our own bootstraps, so to speak, the task becomes much less difficult. This is the dream of Robert Zubrin, who envisions a day when great space tankers ply the void between Earth, Mars, and the outer solar system, fueling a system-wide economy. Nevertheless, there will always be those of indomitable spirit, he believes, who will not be content with even those astounding achievements. "For the best of humanity then, the move must be ever outward," he muses. "It won't end in the vast realm of the outer planets — each step demands another. The farther we go, the farther we will become able to go. Ultimately the outer solar system will simply be a way station toward the vaster universe beyond. For while the stars may be distant, human creativity is infinite."[5]

Notes:

1. Frazer, Lance, "Mars Direct," *Smithsonian Air & Space:* Smithsonian Institution, Washington, DC, April/May 1994, p. 64.
2. Zubrin, Robert, Mars and Luna Direct, *Journal of Practical Applications in Space:* High Frontier, Inc., Arlington, VA, Fall 1992, p. 26.
3. *Ad Astra,* September/October 1992, p. 39.
4. *Ad Astra,* March/April 1993, pp. 19-20.
5. *Ibid.*, p. 23.

Sources:

Zubrin, Robert and Benjamin Adelman, "The Direct Route to Mars," *Final Frontier:* Final Frontier Publishing Co., Minneapolis, MN, July/August 1992, pp. 10-15 and 52-55.

Final Frontier, April 1994, p. 7.

Ten
Forward Unlimited

Dr. Robert L. Forward is deeply concerned about how we are going to get off this planet and move around in space. He is so concerned that he has become a self-styled champion of advanced and alternative space propulsion techniques, including those he invented or improved himself, and those of other space visionaries. He believes it is high time we develop some of these strategies. "NASA needs to stop the interminable paper studies and move into the development and demonstration of advanced forms of space propulsion," he said recently. "That way, mission planners for returning to the Moon, or the exploration of Mars, can have viable alternatives to make such programs economically feasible. Otherwise, this nation is going nowhere in space."[1]

Forward has established a consulting company, called Forward Unlimited, to advocate advanced launch systems. His current favorite is tethers. He envisions combining several independently-invented tether systems into a complete Earth-Moon-Mars transportation system. A space tether is a long filament of some super-strong fiber like Kevlar™ (used in bullet-proof vests) that is set to rotating about a central axis while it orbits a central body such as a planet. A payload attached to one end of the tether at the low point of the rotation can be swung around and thrown off at the high point of the rotation, so that it has momentum that can carry it to a higher orbit, or even to another planet or moon. This can be done because the tether as a unit has its own velocity as it orbits the body, and can impart some of that velocity to a payload.

Of course, for every action there is an equal and opposite reaction, so any velocity imparted to a payload is taken away from the tether, and the tether would correspondingly lose orbital speed and drop in height above the body. This problem can be corrected by sending an equal amount of mass both ways, so that there is no net velocity change in the tether unit, and it will maintain its orbit. For example, in an Earth-Moon tether system, if payloads of, say, lunar dirt are sent down to Earth in equal mass to the material being sent to the Moon, the system experiences no net loss of energy, and needs no outside propulsion once it is put in place (actually, due to drag and friction losses, a little more mass must be sent to Earth, the heavier body, but that is not a large problem).

Forward has combined an Earth tether system invented by Joseph Carroll and a lunar tether system developed by Hans Moravec into an Earth-Moon transportation system. In this scheme, a payload is lifted out of the Earth's atmosphere into a ballistic trajectory by some launch means (due to the strength of Earth's gravity field, there is currently no material strong enough to allow the tether to reach all the way to the surface). There, the payload is captured by a tether and swung into a higher Earth orbit. A second tether in a higher orbit grabs the payload and swings it on a trajectory to the Moon. Upon arrival, the lunar tether captures the payload and deposits it on the surface. The cost would be minimal compared to chemical launchers, for no fuel is required. The cost would only consist of the cost of the payload, a fee for the amortization of the construction of the tether system, and a transport fee. Because of the low costs once such a system is built, several companies operating competing tether systems would strive to offer the lowest-priced delivery service, thereby stimulating the Earth-Moon economy.

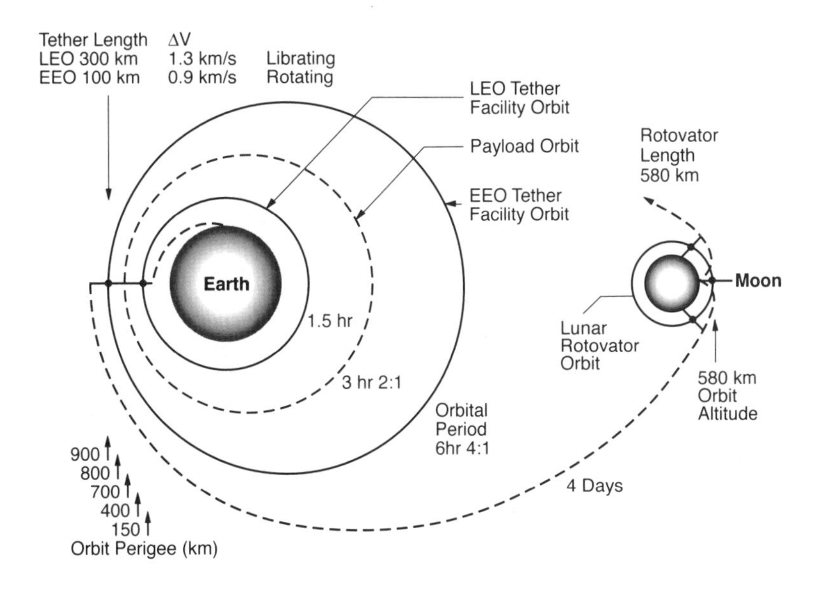

Figure 10-1

Forward Unlimited's conception for an Earth-Moon tether transport system. Designed by Joseph Carroll and Hans Moravec.

The gravitational fields of the Moon or other bodies of similar size, up to the size of Mars, are weak enough that tether systems which Dr. Forward dubs "Rotavators" can be constructed. Either end of a Rotavator can reach all the way to the surface of a body. The lateral motion of an end of a Rotavator as it comes down to meet the surface of a body is almost nil, so payload teams have a good amount of time (around a minute) to detach and attach payloads before it is swept upward again. Elastic connectors at the ends of the tether, which can be reeled in or out, make the system even easier to use. A Rotavator can be set to orbiting a body such that an end of the system meets the surface six times an orbit. As long as the same mass is sent in both directions, the Rotavator will orbit indefinitely, without the need for fuel (other than for drag and friction corrections and other small modifications).

Mars can be added to the inner solar system transportation network by way of a tether system conceived by Paul Penzo. In this scheme, Mars' two Moons, Phobos and Deimos, are used. Tethers are deployed upward and downward from the moons at the same rate, to maintain balance. Since the moons are librationally locked, meaning they always show the same face to Mars, the tethers don't rotate. Once the tethers are fully deployed, a payload can be lifted from Mars by some means (perhaps by Zubrin's NIMF?) to the lower end of the Phobos tether. There it is lifted by electrical means to the moon, and out the opposite tether. The movement of the outward-bound payload along the tether can be employed to generate electrical power, which can be fed back into the system. At the end of the outer tether, the payload is thrown off, and centrifugal force carries it to the height of the lower end of the Deimos tether. There it is once again lifted past the moon to the other end of the tether, where it can be thrown to the Moon or to Earth. A power system is needed to run the Mars tether system, but most of it can be generated by the motions of the payloads along the tethers.

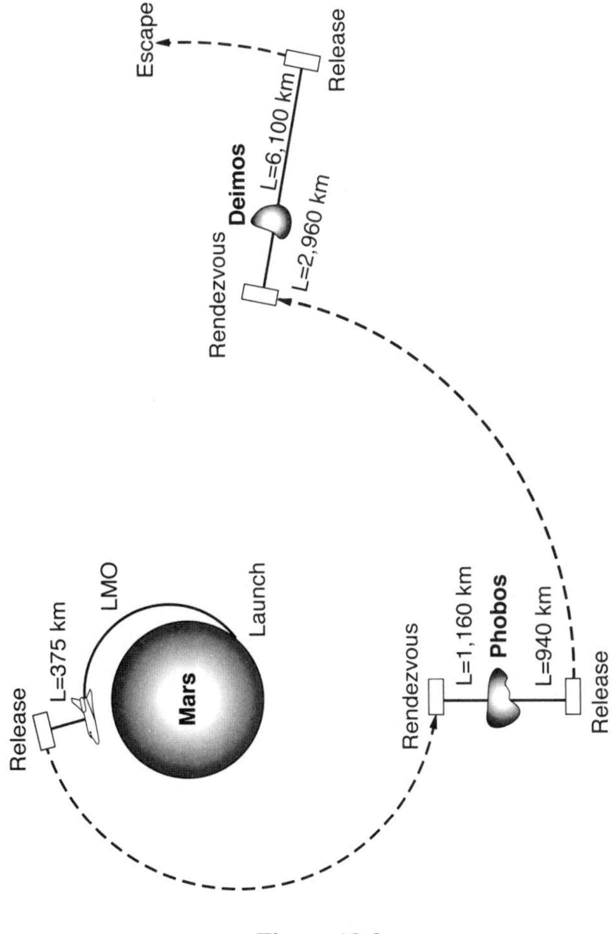

Figure 10-2

Concept for a Mars tether transport system using the planet's two moons, Phobos and Deimos. Designed by Paul Penzo.

Forward envisions a couple of other exotic uses of tethers. One is the cable catapult, which can impart velocities to a payload ten to thirty times greater than a "passive" rotating cable. In this scenario, a large orbiting power plant is coupled to a power-conducting cable. A payload is attached to a linear motor capable of traveling along the cable while getting power from it. The linear motor accelerates along the cable until the payload reaches the desired launch velocity, at which point the payload is released. The linear motor then decelerates to a halt to await the arrival of an incoming payload. This system could toss a payload out as far as Saturn. Tethers could even be set into solar orbits in "bolo" systems, as conceived by Philip Chapman. Bolos would be able to sling payloads around the entire solar system, from bolo to bolo, in orbits of varying diameter around the Sun, inward to Mercury and outward to Pluto. Again, if an equal mass is sent both directions, the system would need no fuel.

Another tether application is the LEO "pumping" tether, conceived by Geoffrey Landis. This tether would connect two masses that would be set into an elliptical LEO. At the perigee (the point in the elliptical orbit closest to Earth) the tether would be contracted. At the apogee (the point in the elliptical orbit farthest from Earth) the tether would be extended. These actions would result in the tether system lifting slightly in its orbit, because the attractive gravity force of the Earth and the repulsive centrifugal force due to orbital motion vary differently with distance, and the difference is enough to impart a net increase in the orbit of the tether. Such a tether and its payloads can "lift itself by its own bootstraps" and travel from LEO to escape trajectory in a month, using only the minimal fuel necessary to cause the contraction and expansion cycles.

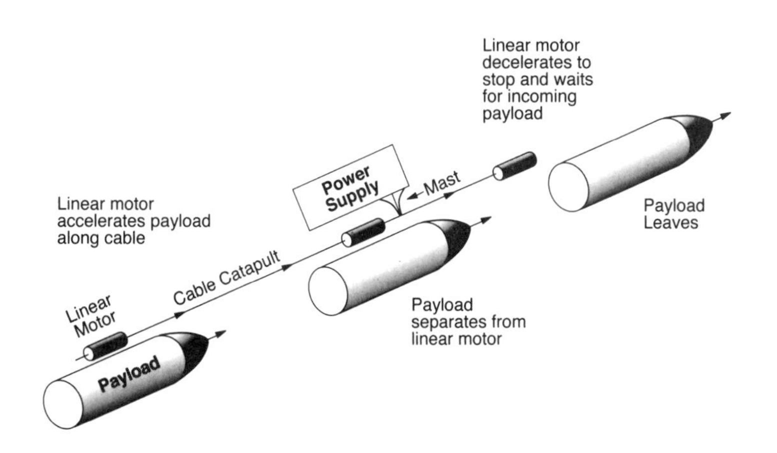

Figure 10-3

A two-way cable catapult shoots a payload into space, then waits for goodies to send back to Earth. Design by Forward Unlimited.

Forward's other big favorite among the advanced propulsion techniques is "mirror matter," or antimatter, propulsion. Far from being a mere *Star Trek* gimmick, antimatter has been collected in significant amounts at the CERN high energy laboratory in Switzerland. An antimatter particle is exactly the same as its normal matter counterpart, except its charge and "spin" are reversed, so that it appears as the mirror image of its normal matter mate, hence "mirror matter." If a mirror matter particle contacts its normal matter mate, both particles are completely destroyed and transformed into energy. This is the most efficient method of converting matter into energy that we have yet encountered.

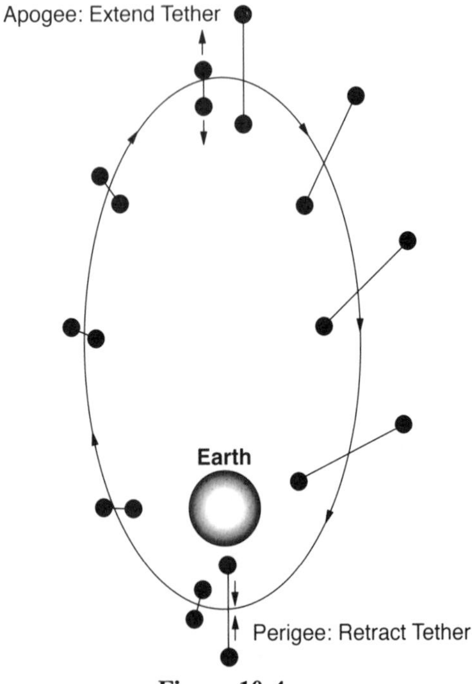

Figure 10-4

Geoffrey Landis' "tether bootstrap propulsion" system uses tether extension & contraction to gradually lift two payloads out of the Earth's orbit.

The problem with mirror matter is that it is rare in our neck of the universe. That is a good thing, for if there were an abundance of mirror matter locally, everything would be exploding all the time. It takes incredible amounts of energy to produce even a tiny amount of mirror matter. In fact, it is a puzzle to the physicists why there is not more mirror matter around. The production of some mirror matter particles theoretically implies that each particle in the universe has a mirror mate. To date, we have not found this missing matter.

That does not stop Forward Unlimited from wanting to use that which we can produce. Current efficiencies of mirror matter production yield about one particle for every billion particles produced in high-energy experiments. However, Dr. Forward asserts, these techniques can be brought out of the lab and into a mirror matter "factory," increasing production efficiencies to one percent of one percent. This ratio still seems dreadfully inefficient, but that 0.01% can be completely converted into energy, along with the particle with which it comes into contact, to produce, as Forward says, 200% efficiency.

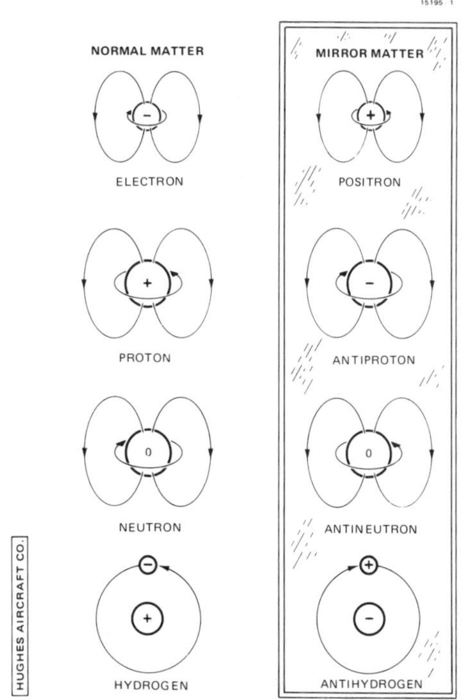

Figure 10-5

By colliding particles of matter with their antimatter mirror images, enormous amounts of energy are created which can be used for propulsion.
(Illustration courtesy of Robert L. Forward and Forward Unlimited.)

Forward estimates that the cost of producing a milligram of mirror matter can be brought as low as $10 million. Furthermore, a 1987 RAND Corporation study concluded that it would take thirty years and $30 billion to build the infrastructure to produce useful amounts of mirror matter. Therefore, it will never be cost-effective to use as a fuel for lifting payloads from Earth's surface. In deep space, however, where fuel must be laboriously and expensively imported or manufactured, it might be just the ticket. A milligram of mirror matter can produce the same amount of energy as twenty tons of the most efficient chemical fuels. Moreover, a lot of the energy in the chemical fuels is applied to moving the as-yet unused portion of the fuel, so less net energy is available for the payload.

For moving around the solar system, Forward has calculated the optimal mass ratio that can be used for any interplanetary ship. That ratio is about two tons of propellant (water, hydrogen, or some similar substance) to one ton of spaceship. The only thing that varies in this design is the amount of mirror matter. For longer or faster journeys, use more mirror matter. For short hops or leisurely cruises, use less. In any case, the amount of mirror matter is negligible in the design calculations. For example, a gram of mirror matter heating twenty tons of water can send ten tons of ship and payload to Mars in a month. Most voyages within the solar system will require much less energy; the amount of mirror matter needed for most missions can be measured in milligrams. The cost of creating antimatter is high, but due to its efficiency in producing energy, it may well be a good long-term solution to interplanetary, even interstellar, travel.

Forward champions several other propulsion technologies. Developed properly and used responsibly, nuclear thermal propulsion generated from fission would eclipse chemical propellants in effectiveness. Even better would be fusion propulsion, but a self-sustaining fusion reaction, which generates more power than it consumes, has not yet been demonstrated.

Laser propulsion is high on Forward's list of promising technologies. The main advantage is that the power supply is not carried on the vehicle, thus avoiding the need to expend energy to lift it along with the payload. This extra energy can be used to lift more payload. In this scenario, worked out in greatest detail by Jordin Kare, a payload would be mounted on a solid block of propellant, such as plastic or water ice. A laser would blast the propellant with two short bursts. The first burst would melt the propellant, forming a thin layer of gas. The second burst would explode this gas, providing thrust against the rest of the solid block. This process would be repeated millions of times at a high rate of speed, making it possible to lift a payload at rest on the ground to a LEO orbit. Steering the payload can be accomplished by changing the point on the propellant block that the laser strikes. For example, to send the payload to the right, the laser is aimed at a point to the left of center of the propellant block.

If speed of travel between planets is not a concern, Forward believes solar sails are the way to go. A solar sail presents a vast, reflective area of around 800 meters in diameter to the light photons streaming from the Sun, just as a sailboat on the water presents a sail surface to Earth's wind. The photons do not provide much thrust, but over time they can impart tremendous changes in velocity. Steering would be accomplished by changing the angle of the solar sail to the stream of photons. To travel in toward the Sun, the sail tilt is adjusted so that the photons decrease the craft's orbital velocity around the Sun, causing it to fall inward. To travel away from the Sun, the sail is tilted to cause the photons to increase the craft's orbital velocity, producing the reverse effect. This technology would be ideal for transporting a steady stream of cargo back to Earth from, say, a mining colony on an asteroid in the main belt between Mars and Jupiter. Once a "pipeline" is established, the long transit times become immaterial, as cargo loads arrive once a month, or some similar frequency.

Forward holds a patent on the use of a solar sail to provide direct broadcast television to the higher latitudes via a "polesitting" spacecraft. Dubbed "Statite," it does not orbit the Earth, but uses the force from the light photons from the Sun to counteract the gravitational attraction of the Earth. Such a craft would "hover" at thirty to one hundred Earth radii above the north or south pole, providing communications services that satellites in Geosynchronous Earth Orbit (GEO) over the equator just cannot deliver, due to their orbital angle.

One day, there may even be solar sailing regattas. The World Space Foundation, headed by Robert Staehle, is designing and building a solar sail for a race around the Moon. Originally Staehle's team was slated to enter a solar sailing competition in 1992 to commemorate the 500th anniversary of Columbus' landing in the new world, but the competition has been postponed until the late 1990s. They will be challenged by solar sailing teams from Europe and Japan. The winner will be the first team that can return images from the far side of the Moon. The competition participants hope this race will mark the beginning of an age when solar sailing ships gracefully ply the vast expanses between the planets.

An even more exotic concept is the magnetic sail, or "magsail", concocted by Robert Zubrin and Dana Andrews. The principle is the same as a solar sail, but instead of the light photons, the magsail employs the solar "wind" of particles streaming from the Sun, which is 1000 times weaker in pressure. This pressure differential is overcome by the size of the magnetic "sail" in relation to a solar sail. In the magsail design, the solar particles press against a magnetic field rather than a physical sail. A vast circle of superconducting wire carries an electrical current, which creates a magnetic field. The solar wind bounces off the magnetic field to produce thrust. Although the diameter of the magsail is much greater than a solar sail, the superconducting wire runs only around the perimeter, whereas the solar sail's diameter is filled with structure. Moreover, the surface of the magnetic field is one hundred times the area enclosed by the actual wire, giving the solar wind a massive area against which to push. This effect gives magsails ten times the efficiency of solar sails. The high-temperature superconducting material necessary for magsails, however, is still at the experimental stage, and is not ready to be produced on the scale necessary to construct interplanetary magsails. We will have to content ourselves with designing and building the less efficient solar sails for now. Yet even those are amazing craft, and will do wonders for interplanetary cargo runs.

When contemplating future propulsion techniques, Dr. Forward does not stop at the edge of the solar system. He has conceived of negative matter drives capable of approaching light speed, "Starwisps" riding a microwave beam at near-relativistic velocities, and other concepts on the edge of feasibility. Most of these ideas end up in his works of science fiction, although the engineers of a few centuries from now might not find the concepts beyond their abilities. We can begin working now, however, on some of the closer-to-home ideas. Forward believes tethers are the next great hope of getting off Earth. He envisions orbital transfer companies offering space tourists competitive deals on transportation to orbiting space vacation facilities (perhaps *Prelude*?). When that day arrives, space will truly be open to the common person.

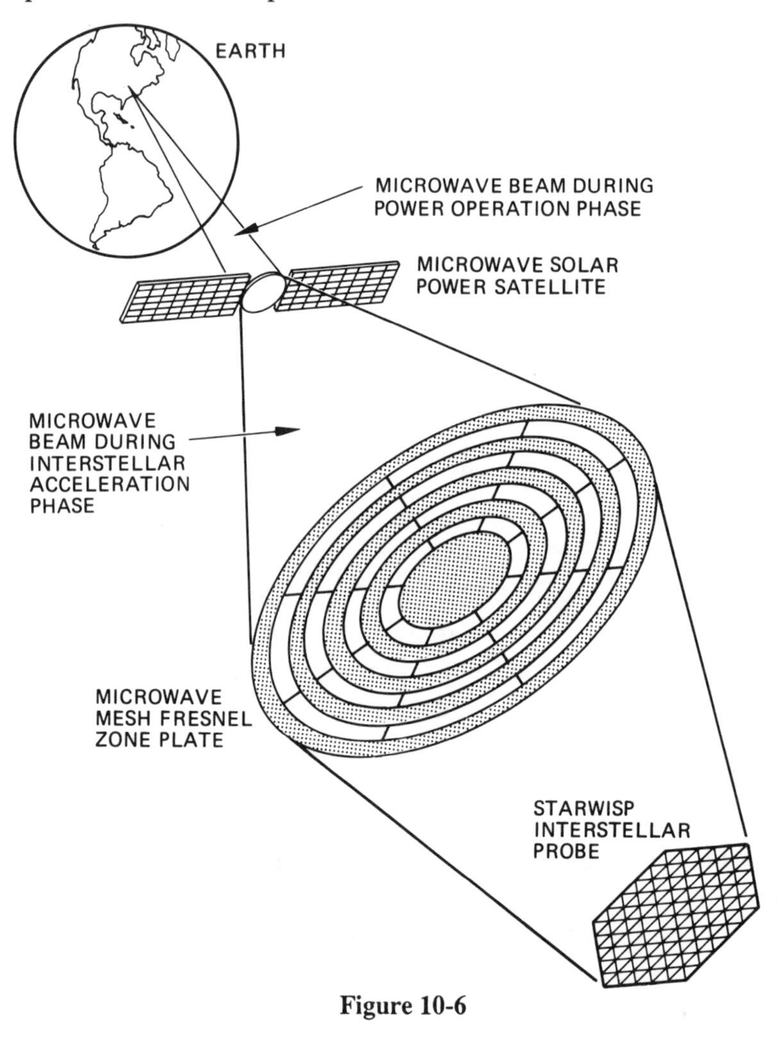

Figure 10-6

The Starwisp interstellar probe combines microwave beams and solar power to explore the galaxy. (Illustration courtesy of Robert L. Forward and Forward Unlimited.)

Notes:

1. *Final Frontier,* March/April 1992, p. 41.

Sources:

Forward, Robert L. and Joel Davis, *Mirror Matter: Pioneering Antimatter Physics,* 1988, John Wiley and Sons, NY.
Forward, Robert L., *Future Magic,* 1988, Avon Books, NY.
Forward, Robert L., "21st Century Space Propulsion," *Journal of Practical Applications in Space,* Winter 1991, pp. 1-35.
Forward, Robert L., "Space Slings," November 16, 1993.
Personal communications, November 16, 1993.

Eleven
The League of the New Worlds

Despite all the excellent efforts chronicled thus far, it still may take a long time before space travel is a common occurrence. This possibility does not frustrate or impede Dennis Chamberland. He intends to design and build the necessary elements for living in space, and be ready to go when the manifest is planned for the first passenger spacecraft. Chamberland founded The League of the New Worlds in 1990 to explore and experiment with the human requirements of living in exotic environments. The plan of the League is to use the ocean as an analog environment to space. Keeping a human alive and healthy under the waves has many of the same challenges as doing so among the stars. The League takes a very hands-on approach to the task. "We are not an organization of enthusiasts, a grass roots movement, or a club," Chamberland proclaims. "We are the explorers. We go, we do, we intend to find and discover. That is our purpose, our only reason for being."[1]

The League is working toward the establishment of the Atlantis Seafloor Colony Project, a habitat which will serve as the nucleus of the first permanent human settlement in the marine environment, and as a laboratory for testing space techniques and processes. Although the League functions primarily in the sea, the space analog is never far from members' minds. Chamberland says he and his fellow aquanauts are "always keeping our focus on the inevitable space connection, always designing for the Moon and Mars."[2]

The first step in that design focus is the Engineering Research Project, or ER-1. This "inner space vehicle"[3], as Chamberland calls it, is a self-contained unit designed to house a two-person aquanaut team. The vehicle is designed for use in the sea or in space. Chamberland says that once a person enters ER-1, that person can no longer distinguish whether he or she is under the sea or in space. The ER-1 may one day be used as the modular basis for a space habitat, but its more immediate purpose is as a platform to study the marine environment, life sciences, and the dynamics of interpersonal interactions in a small, closed space. Such research is essential, Chamberland believes, if we are to successfully settle the space frontier.

Crews have tested the ER-1 in a lake environment, and have already redesigned significant aspects for greater efficiency. This design, test, evaluate, and redesign philosophy, which incrementally improves and evolves the overall system, is a direct result of the League's practical approach. When the design has achieved sufficient maturity, the ER-1 will be anchored on the continental shelf of the eastern seaboard of North America, in approximately 120 feet of water, just below the Gulf Stream. This, according to Chamberland, will be humanity's first true attempt to systematically study the marine environment on a continuous, immersion basis.

The Gulf Stream is the largest single environmental system on the planet, circulating in a large, clockwise pattern from the coast of North America, through the Northern Atlantic, down the European Seaboard, then back across the mid-range of the Atlantic. At any given moment, the volume of 1,000 Mississippi rivers is flowing by a given point. The study of this system will greatly enhance our knowledge of the Earth's ecosystem, and provide us with valuable knowledge we can apply to survival in space. "The imperative [of establishing such a base] is glaringly apparent," Chamberland believes. "Without the human outpost on the seafloor, humanity will be essentially blind to the events occurring on over three quarters of our own planet."[4]

ER-1 will be the first modular unit of The Challenger Station, which will be the nexus of the Atlantis Project. By designing on a modular basis, the League can easily transfer knowledge acquired under the sea to space environments, when the time comes. Design and construction techniques, as well as operations, will not be static, inflexible elements. Rather, they will be adaptable to a wide variety of environments and situations.

The League has already built and tested other elements of the overall colonization system. In 1991 the RD-1-10 was successfully deployed. The RD-1-10, one of many sizes of modular retrieval-delivery (RD) pods, will be used to ferry supplies from the surface to Atlantis. The League is currently designing the ascent-descent (AD) pod, based on the ER-1 vehicle, to be used for transporting crews to and from the station. These elements are all pieces of the puzzle that will eventually be assembled into the fully operational Atlantis seafloor facility. Chamberland estimates that it will take $6 million dollars and three years to get Atlantis firmly established on the seafloor.

Establishing a seafloor colony with a focus on space analogs is a daunting task, Chamberland concedes, but he is convinced that the exploratory nature of humans is equal to the challenge. He points out that in 1909 an advertisement appeared in the *London Times*, seeking a crew for an expedition to Antarctica. The ad read "Men wanted for hazardous journey. Small Wages. Bitter cold. Long months of complete boredom. Constant danger. Safe return doubtful. Honor and recognition in case of success." According to Chamberland, "thousands of men responded to the ad, willing to give up everything for the opportunity to explore the new world of Antarctica."[5] He believes that such spirit still stirs in the hearts of people, waiting to be tapped by the siren call of a new frontier. He has already gathered several stout adventurers into the League, including Mercury astronaut Scott Carpenter, who also logged time in an undersea experiment, and Martin Caidin, an early chronicler of the advance into space and an explorer in his own right. The League's board of directors boasts an impressive collection of Ph.D.s and specialists in several fields, all synergistically working toward the opening of humanity's first true frontier in over a century.

The League is actively preparing its members for their first inevitable steps into space. Through a subsidiary organization called the International Space Academy (ISA), the League offers several courses designed to prepare would-be explorers for the new environments. ISA offers aquanaut and astronaut certification, mission commander certification, and simulated missions including a Moon base (Armstrong Camp), a Mars base (Lowell Base), a Starship voyage (The Dyson), and a seabase (Cousteau Station). When available, ISA also offers to members participation in actual expeditions, including Lake Treks, Ocean Treks, Habitat Operations, and Desert Treks. The new frontier explorers will be well prepared when the day of embarkation arrives.

The Atlantis Seafloor Colony Project will evolve incrementally on a modular basis, with rotating expeditions of ever longer durations. When the time comes, the knowledge gained will be directly transferred to the challenges of establishing permanent human settlements in space. Chamberland is certain such activity is the destiny of humanity. "In any exploration enterprise," he asserts, "it is the *act* of exploration and the *product* of discovery that finally result in economic *return*."[6] With its evolutionary, modular approach to exploration, The League of the New Worlds will take us far along the path of settling the promising new frontiers of the sea and space.

Notes:

1. League of the New Worlds, press kit.
2. League of the New Worlds, press kit.
3. League of the New Worlds, press kit.
4. Robert Heinlein Expeditions prospectus.
5. League of the New Worlds, press kit.
6. *New Worlds Explorer,* League of the New Worlds, 1991.

Sources:

New Worlds Explorer, League of the New Worlds, 1992.
Chamberland, Dennis, personal interview, June 8, 1994.

Twelve
The International Space Exploration and Colonization Company

When the means of traveling to space are finally made available to the general public, the work of colonizing space will begin in earnest. We will need to know how to survive and thrive in space, how to make space our home. That is the goal of The International Space Exploration and Colonization Company, or ISECCo. Headed by Ray R. Collins and based in Fairbanks, Alaska, ISECCo is developing a Closed Ecological Life Support System, or CELSS. Although much more modest than Biosphere II, the work of ISECCo is nevertheless invaluable in the task of opening space to eventual settlement.

The focus of ISECCo's program is Nauvik, a completely enclosed and sealed dome, buried underground, that will be able to provide a modest diet for one individual in a completely closed environment. Nauvik is an Eskimo word meaning "nurturing place." We will need such nurturing if we are ever to truly live in space and cut the umbilical cord to Earth.

ISECCo, a not-for-profit volunteer organization founded in 1988, is building up to the construction of Nauvik through incremental steps. Currently the company manages a garden, in which members experiment with various crops and growing methods, and a minimally enclosed system in Collin's basement, affectionately dubbed the "Basement Biosphere." Recently ISECCo raised enough funds to begin building a partially self-contained greenhouse, which is an intermediate step between the garden and Nauvik. The greenhouse will be useful for testing systems and observing the effects of enclosure on the various participating species, but it will not be entirely closed. Due to heavy carbon dioxide requirements by the plants, the gas will have to be imported, along with water, which will evaporate through the shell. Nevertheless, valuable data will be collected which will transfer directly to the construction of Nauvik.

The construction of the greenhouse began in the summer of 1994 and will be completed by the summer of 1995. Initial trials will be conducted then, and continuous operations will commence during the summer of 1996. The organization is attempting to operate on a shoestring, like true pioneers. Total construction costs are estimated at $17,000, with annual operating expenses of $20,000 or less.

Once a sufficient amount of data is collected and the necessary funds are raised, (estimated at $30,000) ISECCo will begin construction of Nauvik. Nauvik will consist of a forty-foot-diameter dome containing 2.5 floors, which are totally enclosed and buried underground. In the Nauvik design, the lower floor is devoted to large crops, aquaculture (fish), and meat-producing animals, such as chickens or rabbits. The second floor serves as the primary living area. The third floor, which, due to the curve of the dome, only offers half the area of the other floors, is devoted to station-keeping equipment, such as dehumidifiers, water tanks, and air conditioners. Rising through the center of the habitat in a stacked torus (doughnut shape) formation are several small-crop containers, growing short crops like carrots.

The facility is entered through an airlock, which extends laterally from the wall of the dome at the level of the second floor, then turns downward into a pool of water. A person leaving the CELSS climbs down a ladder into the water, swims under the water to the far wall, and climbs a ladder out to the surface. The ladder emerges into a control shack, which assists in the operation of the CELSS. Within Nauvik, there is an emergency escape hatch opposite the airlock.

To research methods employed in CELSS designs and operations, Collins journeyed extensively across the United States in the summer of 1993. He met life support scientists and gave lectures on the state of the ISECCo project at several NASA centers, and even linked up with League of the New Worlds Director Dennis Chamberland in Florida. The similarity of their respective projects led Collins and Chamberland into a synergistic relationship. The two leaders and their organizations now share space colonization ideas and technologies, and support one another's activities. Collins was well received all over the country, from NASA centers to grass roots activist groups working on the problems of opening the space frontier.

The members of ISECCo believe the future of humanity is in space, and that the lessons learned in Nauvik will be instrumental in supporting life off Earth. Nauvik is being designed so that it can be directly translated for use on the Moon or Mars. ISECCo intends to be at the forefront of space development. "In the short time ISECCo had been in existence," Collins writes, "we have grown from a small group of dedicated enthusiasts to a strong international organization. The greatest emigration ever is about to begin. Over the next few centuries civilization will spread throughout the solar system and beyond. Join ISECCo and be in the vanguard of the greatest movement mankind has ever undertaken."[1]

Notes:

1. ISECCo newsletter, 1993.

Thirteen
Where Do We Go From Here?

There you have it. You have encountered several fantastic, excellent plans to get to space. They are not idle fantasies; they are doable, and the people involved are doing them right now.

None of them, however, are complete in their own right. And there are many more plans and schemes and dreams out there, to get us off the planet. We need all of them, if we are to truly colonize the planets and shoot off to the stars.

We now know that the massive, monopolistic, monolithic Beast that is NASA is not going to get us, the average people, off the planet in our lifetimes. Yet we must not replace one Beast with another. We must avoid the scenario so aptly stated by Pete Townshend: "Meet the new boss; same as the old boss." Armed with the knowledge in this book, however, "We won't get fooled again."[1]

The space movement requires a massive parallel approach, in which several activities are carried out at the same time.

Instead of "The Space Program," we need a multitude of space projects. These projects can be linked together into ad hoc, modular arrangements to achieve specific goals, in what I term the "tinkertoy" approach (an unglamorous yet highly descriptive label). When the goal is achieved, the project group is disassembled, but the pieces remain, to be reassembled into a new configuration for the next project.

These projects should never be institutionalized, for that is the beginning of the monolith. Each element of the "tinkertoy" set retains its independence, and can move on to new, promising applications of its capabilities once it has concluded its relationships in the current project.

This modular approach allows for great flexibility and responsiveness to new information, which we shall surely harvest in plenitude as we advance outward. It also does not close off access to newer "tinkertoy" elements; if someone comes up with a great idea that fills a heretofore unmet need, they can develop it and attach it to a "tinkertoy" project. No Beast will stomp it out of existence.

The "tinkertoy" approach also mitigates disasters and catastrophes. If one or another element is taken out because of some misfortune, another element can replace it, and we can keep moving forward. We avoid grounding the whole space effort, as happened after the Challenger accident.

Finally, it allows for the greatest latitude in personal expression. Nobody knows what the people want more than the people themselves. George Orwell looked into the future and saw Big Brother. The future did not transpire exactly as he foretold, but the seeds are there. Yet the modularity and the popularity of such "tinkertoy" elements as video cameras and computers connected by the Internet have enabled people to communicate, even when the government did not want them to, such as during the Tiannanman Square incident, or the recent Russian parliament struggle.

The people want to explore and colonize space, and the governments of the world can stand in the way only so long. History is replete with the end runs the people have pulled on their governments, and space will be no exception. Those in power would do well to join us, rather than block us, or they may be left stuck on the ground, watching us wave good-bye as we blast off into the future.

Notes:

1. The Who, "Won't Get Fooled Again," *Who's Next,* MCA, 1971.

Appendix
Businesses, Publications and Organizations

Following are the addresses of businesses covered in this book, plus other organizations working to open the space frontier.

International Space Enterprises
4909 Murphy Canyon Rd., Suite 330
San Diego, CA 92123

LunaCorp
4350 N. Fairfax Dr., Suite 900
Arlington, VA 22203

The Lunar Resources Company
P.O. Box 590213
Houston, TX 77259-0213
Sponsors *The Artemis Project*

OUSPADEV, The Outer Space Development Company
1400 E. Oakland Park Blvd., Suite 206
Fort Lauderdale, FL 33334
Outernet computer bulletin board number (after 6 p.m. eastern standard time): (305) 564-0089
Publishes *Outward*, a quarterly newsletter on commercial space development.

Kistler Aerospace Corporation
3760 Carillon Point
Kirkland, WA 98033

Hudson Engineering
P.O. Box 2500
Menlo Park, CA 94025

Robert Zubrin
Martin Marietta Astronautics
P.O. Box 179
Denver, CO 80201

Forward Unlimited
P.O. Box 2783
Malibu, CA 90265-7783

League of the New Worlds
P.O. Box 542327
Merritt Island, FL 32954

ISECCo., The International Space Exploration and Colonization Company
P.O. Box 60885
Fairbanks, AK 99706

National Space Society
922 Pennsylvania Ave. SE
Washington, DC 20003-2140
Publishes *Ad Astra*, a bi-monthly magazine on general space developments.

Lunar Reclamation Society, Inc.
P.O. Box 2102
Milwaukee, WI 53201-2102
Publishes *Moon Miners' Manifesto*, a monthly newsletter on alternative space development concepts.

Space Studies Institute
P.O. Box 82
Princeton, NJ 08542
Organization to achieve Dr. Gerard K. O'Neill's space vision, as described in the book *The High Frontier.*

Space Frontier Foundation
16 First Ave.
Nyack, NY 10960
Publishes *Space Front*, a quarterly newsletter on space activist activities.

First Millennial Foundation
P.O. Box 347
201 Railroad Ave., Suite 100
Rifle, CO 81650
Organization to actualize the space vision detailed in the book *The Millennial Project.*

The House of Tomorrow
P.O. Box 801
Greeley, CO 80632-0801
Organization to promote SSTO vehicles
e-mail: sstoHOT@col.com

Houston Space Society
P.O. Box 266151
Houston, TX 77207-6151
Publishes *Journal for Space Development*, covering alternative space development concepts.

World Space Foundation
P.O. Box Y
South Pasadena, CA 91031-1000
Organization to build solar sailing craft.

SUNSAT Energy Council
c/o ETM Inc.
P.O. Box 67
Endicott, NY 13761
Organization to promote solar power satellites.

Publications

Final Frontier
1516 W. Lake St., Suite 102
Minneapolis, MN 55408

Bi-monthly magazine featuring all aspects of space development. Recently improved commercial space business reporting. Also sponsors *Space Explorers' Network*, which features a computer bulletin board. Subscription: $17.95/year.

Space News
6883 Commercial Dr.
Springfield, VA 22158-5803

Weekly newspaper covering all business aspects of space development. Excellent resource. Subcription: $89/year.

Journal of Practical Applications in Space
2800 Shirlington Rd., Suite 405-A
Arlington, VA 22206

Quarterly journal covering alternative methods of space access and development. Features different authors each issue. Subscription: $30/year.

Countdown
P.O. Box 9331
Grand Rapids, MI 49509-0331

Bi-monthly newsletter that covers space activities around the world. Includes "Space Available," a column on space business and investing, written by the author of this book.

Index